I0286671

Charles Dixon

The Migration of British Birds

Including their Post-Glacial Emigrations as Traced by the Application of a New Law of Dispersal

Charles Dixon

The Migration of British Birds
Including their Post-Glacial Emigrations as Traced by the Application of a New Law of Dispersal

ISBN/EAN: 9783337272586

Printed in Europe, USA, Canada, Australia, Japan

Cover: Foto ©berggeist007 / pixelio.de

THE MIGRATION

OF

BRITISH BIRDS

INCLUDING

Their Post-Glacial Emigrations as Traced by the Application of a New Law of Dispersal

BEING

A CONTRIBUTION TO THE STUDY OF MIGRATION, GEOGRAPHICAL DISTRIBUTION, AND INSULAR FAUNAS

By CHARLES DIXON

AUTHOR OF "THE MIGRATION OF BIRDS," ETC., ETC., ETC.

WITH SIX MAPS

LONDON: CHAPMAN AND HALL, Ld.

1895

Richard Clay & Sons, Limited,
London & Bungay.

PREFACE

The present volume is the result of much additional thought and hard work generally upon the subject of Avian Emigration and Migration. It is a subject that has a peculiar fascination for me, doubtless because it offers all the charm of novelty, and is as yet a practically unworked field of research. No branch of Ornithology can possibly be more interesting than that which treats of the Dispersal—either by Emigration or Migration—of Birds over the globe.

To many naturalists this dispersal may appear entirely fortuitous, or very largely due to such abnormal influences as Glacial Epochs. I honestly confess that for many years I was imbued with very similar ideas—what all men accepted as true must assuredly have been truth. But after a long and careful study of the phenomenon of Emigration (or Range Extension) I began to doubt some of the most generally accepted views respecting the Geographical Distribution of Species. Difficulty after difficulty arose, convincing me more and more that the solution of the problem must be sought in other directions. After much hard work at the distribution of Birds generally, I have been enabled to propound what I believe to be a Law of Dispersal. The present volume is the development and application of that Law. I may here remark that this Law has been gradually developed from a vast number of accumulated facts, and is not a result of any preconceived theory.

I have selected British Birds for this application, chiefly because our knowledge of their distribution, and of the past physical changes in the areas they inhabit, is not only extensive but fairly reliable. We are often told that there is nothing new to be said about British Birds. I offer the present volume as an answer to that hackneyed remark, and desire specially to call attention to our lack of information on many important subjects connected with British ornithology, indicated here and there in the following pages, as a possible field of very fruitful research.

I am aware that much in the present volume is opposed to the general views held by naturalists, and even to those expressed by myself in former works. For such portion of my work, a patient perusal and an impartial judgment are asked. I am quite prepared to meet with some amount of hostility from specialists whose views are not in harmony with this new Law of Dispersal. I have no fear for the results so far as Birds are concerned; and I await with profound interest, and yet with every confidence, the results of its application to other branches of biology, to which only specialists are competent to apply it. On its important bearing on the Distribution of Floras I have already dwelt at some length in my closing chapter.

For the sake of convenience the work has been divided into two parts; the first treating of physical and climatic changes, and their effects on Birds and species generally; the ancient Emigration of birds to the British Area, and its modern progress; lastly of Island Avifaunas and their bearing on the entire subject. The second part deals with the Migration of Birds within the British Area; the Routes those birds follow; the conditions of their Flight; the spring and autumn aspects of the phenomenon; and lastly, with the very curious and interesting subjects of Local Movement or Internal Migration, and Irruptic Movement.

I earnestly hope that the present work may assist in solving some of the many intricate problems connected with the present and past Geographical Dispersal of Life, and the Migration of Birds; for then its purpose will be amply attained, and the labour involved in its construction ever recalled with pleasure.

CHARLES DIXON.

CONTENTS

PART I.—DISPERSAL:

A STUDY OF THE PHENOMENON OF AVIAN EMIGRATION TO THE BRITISH ARCHIPELAGO AND ADJOINING AREAS.

CHAPTER I.

PAST GEOGRAPHICAL MUTATIONS AND CLIMATIC CHANGES.

CHAPTER II.

RANGE BASES OR REFUGE AREAS.

CHAPTER III.

THE GLACIAL RANGE CONTRACTION AND POST-GLACIAL EMIGRATION OF BRITISH BIRDS.

CHAPTER V.

RECENT EMIGRATION.

CHAPTER VI.

ISLAND AVIFAUNAS.

PART II.—MIGRATION:

A STUDY OF THE PHENOMENON OF AVIAN SEASON-FLIGHT ACROSS THE BRITISH ARCHIPELAGO.

CHAPTER VII.

ROUTES OF MIGRATION.

CHAPTER VIII.

CONDITIONS OF FLIGHT.

PAGE

Routes of Migration, how followed by Birds—Paley's Definition of Instinct—Impulse of Migration—Restlessness of Captive Birds—Certain Routes followed by Certain Individuals—How a Route of Migration has been Learnt—Mysterious "Sense of Direction" a Myth—Altitude of Migration Flight—Advantages of a Lofty Course—The Order of Migration—A few Old Birds migrate as Early as the Young—The Daily Time of Migration—Amount of Sociability amongst Birds on Passage—The Perils of Migration 234

CHAPTER IX.

THE SPRING ASPECTS OF MIGRATION IN THE BRITISH AREA.

Commencement of Spring Migration in the British Islands—Departure of Eastern Migrants—Departure of North-Eastern Migrants—Abnormal Lines of Migration from the British Area—Birds migrating too Early—Arrival of First Spring Migrants from the South—Departure of Winter Visitors to the British Islands—Coasting Migration in Spring—Migration of various Northern Birds—Arrival of Summer Migrants in the British Islands—The Growing Intensity of Spring Migration—Months of Passage of various Species—Gradual Advance northwards of Migrants—Duration of Spring Migration—Vertical Migration in Spring—The various Species performing it—Order of Migration—Table indicating the Spring Migration of Birds across the British Islands 243

CHAPTER X.

THE AUTUMN ASPECTS OF MIGRATION IN THE BRITISH AREA.

Migration more Apparent in Autumn than in Spring—Difficulties of Observing the Phenomenon—Commencement of Autumn Migration in the British Islands—Arrival of Birds from the North and North-East—The Species that are the

PAGE

Earliest to Arrive—Growing Intensity of the Movement—The Hardier Species are Latest to Appear—Autumn Migration of Fieldfare and Redwing—The Earliest Departures from the British Area—Early Migrants Abnormal—The Growing Intensity of Southern Migration as Autumn advances—Duration of Migration Periods—The Migration into the British Area from the East—General Aspects of the Phenomenon—Abnormal Lines of Migration in Autumn—Cross Migration in Autumn — Reversal of Route by Migratory Birds—Erroneous Interpretation of the Facts—The True Explanation—Duration of Autumn Migration—Vertical Migration in Autumn—Order of Migration—Table indicating the Autumn Migration of Birds across the British Islands 260

CHAPTER XI.

INTERNAL MIGRATIONS AND LOCAL MOVEMENTS IN THE BRITISH ARCHIPELAGO.

Meagreness of Data bearing on this Question—Local Movements in Spring and Summer—Internal Migration always takes place within Normal Areas of Dispersal—Birds do not wander from their Areas—Comparative Movements of the Snow Bunting, the Northern Bullfinch, and the Crested Titmouse—Emigration only undertaken during the Season of Reproduction—Local Movement amenable to Law—Effects of Severe Winters on Birds—Results of such Movements ineffectual in extending Area—Redwings and Severe Weather—The Want of Carefully-kept Records—Local Movements at Lighthouses—Irruptic Movements—Their Futility as Colonizing Agents 276

CHAPTER XII.

SUMMARY AND CONCLUSION.

The Present Volume illustrates the Development and Application of a New Law of Dispersal—Past Geographical and Climatic Changes—The Glacial Epoch and its Bearing on the New Law of Dispersal—Effects of the Glacial Epoch on Species—Range Bases—Application of the Law of Dispersal to the Range Contraction and Emigration of British Birds—The Migration of British Birds—Routes of Migra-

LIST OF MAPS.

THE
MIGRATION OF BRITISH BIRDS

PART I.—DISPERSAL

A STUDY OF THE PHENOMENON OF AVIAN EMIGRATION
TO THE BRITISH ARCHIPELAGO AND ADJOINING AREAS

B

PART I.—DISPERSAL.

CHAPTER I.

PAST GEOGRAPHICAL MUTATIONS AND CLIMATIC CHANGES.

British Fossil Birds—Definition of the Present Work—The British Area—West European Submarine Plateau—The British Seas—Ancient Land Areas between Greenland and Europe—Effects of Submergence—The great Submergence of the British Islands—Evidence against it—Physical Changes during third Glacial Period—Geography of Western Europe during late Pliocene, Pleistocene, and early Post-Glacial Time—Commingling of Pliocene and Pleistocene Species—Glaciation correlated with Elevation and Subsidence—Changes in the Earth's Centre of Gravity—Coast-line of British Area at the close of the Glacial Epoch—Geographical and Climatal Conditions—The 50-Fathom Contour—The 40-Fathom Contour—The 20-Fathom Contour—The 15-Fathom Contour—Commingling of Southern and Temperate Species—Absence of large Southern Mammals from European Post-Glacial Deposits—The Flora and Fauna of Greenland and Iceland—Post-Glacial Changes of Climate—Effects of Physical and Climatal Change on Birds.

OWING to the excessive rarity of fossil bird-remains, we possess but the scantiest palæontological evidence of that avifauna which occupied the British area even during Pleistocene and late Pliocene time, but judging from the fragments hitherto discovered it presented very similar aspects to that existing now in our islands. With regard to Miocene and Eocene time it is still more difficult, from the present state of our knowledge, to derive even a general idea of its aspects. Of this, however, we may rest assured, that it possessed many dominant features which have long been entirely obliterated from existing types. Amongst these vanished birds we may recall the giant *Gastornis klaasseni;* the Accipitrine *Lithornis vulturinus;* the imposing *Argillornis* and *Odontopteryx*—allied to the Gannets and the Cormorants; *Halcyornis* and *Proherodius*, representing our modern Gulls and Herons; the Flamingo-like *Helornis*, and the Ibis-like *Ibidopsis*. We need not, however, speculate on the avifauna of so remote a past; it has little or nothing to do with the past emigrations of existing species, or of the phenomenon of Migration across what is now the British Archipelago, the philosophy of those grand avian movements being amply demonstrated by the forms inhabiting that area during present time.

For the purposes of the present volume it will not be necessary to go back, geologically, to a very distant past. On the ancient origin of Avian Season Flight I have already dwelt at some length in the *Migration of Birds*. The present work is more a study of the Post-Glacial Emigrations of our avifauna and of Migration as undertaken by birds in our own day, going on unceasingly around us from year to year, than a history of the phenomenon of Migration in remoter ages. Nevertheless we shall find in our investigation of the interesting subject that it will be advisable not only to go back at least to late Pliocene and to Pleistocene or Quaternary time, but to include the avifaunas of many areas more or less

100 & 50 FATHOM CONTOURS OF THE BRITISH AREA

Migration of British Birds Plate I

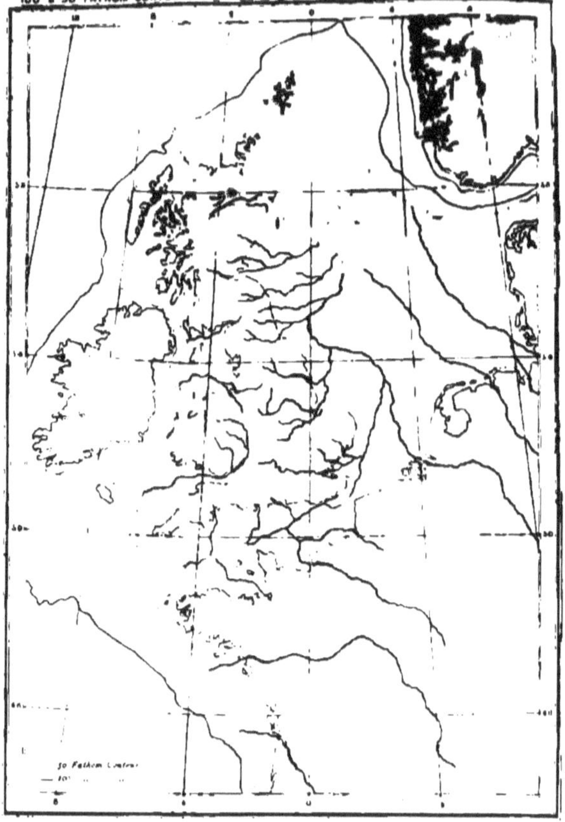

CHAPMAN & HALL, LIMITED

contingent to our own, in order to arrive at a reasonable and probably correct interpretation of the facts now presented. It is absolutely impossible to understand the present migration and geographical distribution of British birds, unless we take into consideration the past emigrations of these species, or their common ancestral forms; for undoubtedly those ancient emigrations present the only means now available by which we can explain a very large number of apparently anomalous facts. It is equally impossible to understand them without taking into consideration the various geographical mutations and climatic changes which have taken place in this area during late Tertiary and Post-Tertiary time. It becomes necessary therefore to glance briefly at the probable state of Europe, not so much during the height of the Glacial Epoch, but more especially during the closing periods of that awful era and in the earlier stages of Post-Glacial time.

The area which now forms the British Islands consists of the elevated portions of a vast submarine plateau, receding from the coasts of continental Europe, and for the most part covered by a shallow sea varying from fifteen to a hundred fathoms in depth. This plateau extends from Denmark, some distance beyond the south coast of Norway, the Shetlands, the west coast of Ireland, and southwards along the French coasts of the Bay of Biscay. Beyond this limit westwards the sea rapidly deepens in a series of terraces from 500 and 1000 to 2000 fathoms. The greater part of this water area, including most of the North Sea, ranges from 15 to 50 fathoms, but deeper trenches occur along the centre of the bed of St. Georges Channel and the

Irish Sea, in the North Channel between Ireland and Scotland, and between the Inner and Outer Hebrides; these hollows range from 100 to 150 fathoms in depth, and may be well described as a series of submarine lakes and ancient river channels. Another narrow deep channel, which was apparently an ancient fjord or lake, extends round the southern coast of Norway, and is for the most part some 500 fathoms in depth. To the north-west of the Orkneys a narrow ridge, under 500 fathoms from the surface, extends to the bank of less than 300 fathoms reaching to Greenland, and on which the Faroes and Iceland are situated. It is probable that this north-western and comparatively shallow submarine bank was formerly a more continuous land area, if not, as some authorities suggest, an actual isthmus, connecting Greenland with continental Europe even during early Post-Glacial time. It is interesting to remark the vast effect such an isthmus would produce upon the climate of Northern Europe by shutting out the warm waters of the Gulf Stream, which would unquestionably reduce the climate of Scandinavia to the same state as that now prevailing in Greenland, and clothe many of the Scottish mountains in canopies of perpetual snow. The submergence of this land connection has been the chief means of rendering the climate of North-western Europe so comparatively mild, by admitting warm currents to circle round the coasts and exert their beneficial influence on fauna and flora alike to a most remarkable extent. The effect on the Migration of birds alone of such a physical change has been far-reaching and profound, as may be readily perceived.

The accompanying map will better illustrate the

probable appearance of Western Europe during at any rate the closing period of Pleistocene and early Post-Glacial time. That a large portion of the British Islands has been submerged to a depth of about 1400 feet during the Glacial Epoch has long been the almost universal opinion of geologists; the chief, if not only evidence of this great submergence being the occurrence in a few localities of glacial gravels and drifts containing marine shells. Strong evidence against this "great submergence," however, is to be found in the fact that the shells in these drifts differ considerably in habit, some belonging to sandy, others to muddy bottoms, some only known to exist below tidal water, others confined to the shore, consequently they could not well have lived together in the places where their for the most part fragmentary remains are now found. Again, no evidence of this submergence is to be met with on the English lowlands, which, if it had really occurred, would have formed the bottom of a sea no less than 1200 feet in depth, and must have left traces of old beaches or marine remains characteristic of such a vast depression. As Dr. Wallace recently remarked (*Fortnightly Review*, Nov. 1893, p. 633): "In consequence of these various difficulties it was suggested by the late Mr. Belt that the great Irish Sea ice-sheet had carried up a portion of the sea-bottom embedded in its substance, perhaps containing deposits of shells of various periods, and thus explaining the intermixture of species as well as their fragmentary condition. The fact that boulders and pebbles from Scotland, Ailsa Craig, and Cumberland have been found in the Moel Tryfaen beds almost amounts to a proof that they were so uplifted; and a recent search has shown that in the other localities where marine shells have been found in drift at great elevations similar foreign rocks occur, rendering it almost certain that the same ice-sheets which have distributed foreign erratics so widely over our country, and which in doing so *must* have passed over the sea-bottom, have in a few cases carried with them a portion of that sea-bottom and deposited it with the erratics in the places where both are now found." If the drift be taken as evidence of submergence, then, as Mr. P. F. Kendal pointedly remarks (*Man and the Glacial Period*, p. 178): "A subsidence of the Yorkshire Wolds took place on the east, but not in the centre or west; that the Pennine Chain was submerged on the western side to a depth of 1400 feet, and on the east to not more than 300 feet, even on opposite sides of the same individual hill; that all the lowlands between, say, Bacup and the Welsh border, were submerged, and that the hills near Frondeg partook of this movement, but only on their eastern sides; that the centre of Wales was exempt, but the summit of Moel Tryfaen forms an isolated spot submerged, while the surrounding country escaped. These absurdities might be indefinitely multiplied, and they must follow unless it be admitted that the phenomena are the results of glacial ice, and that the ice can move uphill." That the latter circumstance is a fact I need scarcely say has been proved by irrefutable evidence. It is also very interesting and suggestive to remark how the supporters of the "great submergence" hypothesis assert that the subsidence dwindled down to zero in the southern portions of England, say to a line drawn between Bristol and London, curving northward as far as Warwick, which, it may be remarked, is the ascertained limit of glaciation, the area below that boundary, embracing all the English shires south of the Bristol Channel, and the valley of the Thames, having for the most part endured from the early Pliocene Period. It is also significant that no drift containing marine shells has been found at high levels in Scotland.

Whatever changes may have taken place elsewhere in British Europe during the Glacial Epoch, it seems pretty certain that the land which now lies buried under portions of the Bay of Biscay down to the borders of the French Landes, beneath the English Channel, the Bristol Channel, St. Georges Channel, and to the south of Ireland, endured through most if not all of that period, and formed the bridge by means of which our area was invaded by the large mammalia and other forms of animal and vegetable life during the successive mild inter-glacial periods. We shall yet see the even greater importance of this southern land mass. It is, however, suggested that during the third glacial period the sea advanced up the English Channel—the high temperature of its waters probably checking the southern advance of the ice-sheets beyond the valley of the Thames. Whether the depression extended up the Bristol Channel or the Irish Sea area need not concern us, for we have evidence to suggest that that region was entirely glaciated or snow-clad. It is presumed that during this era the Selsey beds and the various raised beaches which occur from Cornwall to Brighton on the English coast, and from Brittany to St. Valèry sur Somme on the French coast were formed. During this glacial period the North Sea is said once more to have made its appearance, and a portion of its southern coast-line extended far within the present limits of Northern France and Belgium. It is during this era that the range of the Greenland whale and the walrus is reputed to have extended as far south as the coast of Lincolnshire. This state of things probably endured long enough for the waters of the two seas to meet across the narrow isthmus which joined England to France, but was followed by terrestrial upheaval or marine depression, and as the glacial period passed away the Channel and the North Sea once more became dry land.

It will perhaps be most convenient to deal first with the geographical mutations which our area has undergone during the last phase of Pliocene time, and from the close of the Glacial Epoch down to the period when the British Islands were finally severed from continental land, so that we may better understand the various points at issue, and be able to refer with clearness to such changes whenever the exigencies of our investigations render such a course advisable or necessary. We have already seen (see Map, p. 5) that a vast submarine plateau or bank covered by a shallow sea extends far into the Atlantic, some 50 miles beyond the British area to the north and west, and some 150 or 200 miles to the south off Finisterre, whence it is contracted down the shores of the Bay of Biscay to Biarritz. Upon this submarine plain many important physical changes have occurred since the close of the Pliocene Period. We possess sufficient evidence to suggest that in those remote ages the North Sea was much more contracted, and the land mass of our area was very much more extensive than it is during present time. From the Shetlands to the Bay of Biscay, following the 100-fathom line, was apparently dry land, our eastern coast-line conforming much to the general direction which it does now, extending, however, some 70 miles further out to sea than in our time, to as far south as the Wash, where, however, the sea encroached upon what is now land in North-east Norfolk and East Suffolk and Essex to a little north of the Thames. The continental coast was equally extended, the shores of the Netherlands approaching our own to within some 70 miles at the narrowest part. During this late Pliocene Period, therefore, a great portion of the southern part of the North Sea was a wide alluvial plain, studded with lakes and intersected by sluggish streams, very similar in character to the Broad district of to-day, bounded on the west by higher forest-clothed ground, and traversed by a continuation of the Rhine with the Thames as a tributary stream. Of the fauna and flora of these remote days, according to Professor J. Geikie (*Prehistoric Europe*, pp. 261, 334), the remains include "those of elephant, hippopotamus, horse, cave-bear, urus, Irish deer, and many other Cervidæ. The fauna is remarkable as showing commingling of Pliocene and Pleistocene species. Thus we have among the former a bear (*Ursus arvernensis*), a rhinoceros (*R. etruscus*), and a deer (*Cervus polignacus*) which have not yet been met with in any deposits of more recent age. Again, several of the forms which appear in the 'forest bed' are common Pliocene species that seem to have vanished from the European fauna in early Pleistocene times. Among these are *Machairodus*, and others which occur in the older Pleistocene deposits, but have not been dug up in beds pertaining to the latest stage of the Pleistocene Period. Nevertheless, the characteristic Pleistocene fauna is well represented in the 'forest bed' of Norfolk by such animals as cave-bear, wolf, fox, wild-boar, urus, mammoth, Irish deer, roe, stag, beaver, and mole. The fauna of the 'forest bed' is thus intermediate between that of the Pliocene on the one hand, and the Pleistocene on the other, and is more closely allied to the latter than the former." Professor Geikie continues: "Underneath the oldest known boulder clay—that of Cromer in Norfolk—occur certain fluvio-marine deposits, the plant remains in which bespeak the kind of climate that characterized England at the commencement of the first glacial epoch. That flora embraced Scots fir, spruce fir, yew, alder, oak, birch, white and yellow water-lilies, bog-bean, common sloe, etc.—indicating a climate perhaps a little colder, but not essentially differing from that of Norfolk at present. But as the plants are traced upwards through the strata they were found by Mr. Nathorst to become more and more stunted and meagre, until in a bed immediately underlying the boulder clay he came upon the Arctic willow (*Salix polaris*) and a moss (*Hypnum turgescens*) now confined in temperate latitudes to the highest alps."

We possess but little reliable information respecting the physical history of the British area during the Glacial Epoch, and great difference of opinion prevails among authorities thereon. Fortunately such history does not very intimately concern the subject of our investigations, which in its entirety is amply accounted for by glaciation alone, or embraced by Post-Glacial time. Of one thing we may be pretty certain, that when the penultimate inter-glacial period had attained its meridian the climatal conditions ("more humid, and much more equable") of our area, and the fauna and flora inhabiting it, resembled very closely those which characterized later Pliocene time. It seems probable, however, that the shallow North Sea of late Pliocene ages had vanished, perhaps partly due to upheaval, especially in the north, or marine depression, but more likely, to a very great extent, silted up by the vast quantities of detritus carried by the Scandinavian and Scottish ice-sheets that several times swept its bed. For, as Mr. Jukes-Brown remarks (*The Building of the British Isles*, p. 426): "a recent boring at Utrecht has proved that the surface of the Pliocene deposits is more than 500 feet below that part of Holland, and as the Pliocene sea probably deepened northward, it is therefore hardly too much to assume that the same surface lies some 100 fathoms below the present bed of the North Sea between Norway and Britain, and consequently that the upheaval necessary to convert this sea into land after the Glacial Period was 100 fathoms less than would have been required to effect the same result in later Pliocene time." With the coming on of the third glacial period, the "great submergence" of the British Islands is presumed to have occurred; but, as we have already seen, this great subsidence could never have taken place—and Professor Geikie himself admits that there is no evidence of any such universal depression in Scandinavia, unless all trace has been removed by succeeding glacial action, which does not seem probable. It is suggested, however, that during this epoch the North Sea appeared once more, and that the Atlantic invaded the English Channel. The low grounds of Prussia are said also to have been submerged about this time.

There can be no doubt that periods of glaciation have always been accompanied by great variation in the relative level of sea and land, but whether so much of this has been caused by elevation or subsidence of vast areas of land, as many geologists so positively assert, seems to the present writer somewhat doubtful. I am disposed to agree with the late Dr. Croll, who has conclusively shown, almost to the extent of absolute demonstration, that oscillations of sea-level may be produced by an alteration of the earth's centre of gravity, caused by the greater amount of ice at one pole than the other, the ocean conforming to such alteration, and appearing to rise in one hemisphere and to fall in the other. Thus the sea-level might rise in a given area without any terrestrial movement whatever. Dr. Croll points out that the great ice masses which were characteristic of the last Glacial Epoch would cause a rise in the sea-level by altering the earth's centre of gravity, and he suggests (according to Professor Geikie) "that some of our recent raised beaches may indicate periods when the ice of north polar regions attained a considerable augmentation, while, at the same time, the ice of the Antipodes suffered a corresponding diminution." The enormous ice-sheets which once covered so much of the Northern Hemisphere may therefore have caused sufficient change in the earth's centre of gravity to produce an oscillation in the sea-level of several hundreds of feet irrespective of any land movement whatever.

To whatever cause, however, we may ascribe it, the British area at the close of the Glacial Epoch appeared to have stood at least 300 feet higher than it does at the present time, or the sea was depressed to that extent. The coast-line of the British area at that period would, generally speaking, correspond at least with the contour of 50 fathoms, but the land surface may possibly have been even more extensive. Indeed such eminent authorities as Lyell, Godwin-Austen, and Professor Geikie fix the coast-line of this period at the 100-fathom contour, which would go far to suggest that the land area of this portion of Western Europe was as extensive at the close of the Glacial Epoch as during the later stages of that era. As Professor Geikie remarks: "In short there was a return of those geographical conditions which we have every reason to believe characterized certain Inter-glacial epochs." It seems much more reasonable to suppose, however, that these geographical conditions had endured to a very great extent, as we have already seen that the great submergence hypothesis cannot be supported by any reliable evidence. Whatever was the exact contour, we may be pretty certain that the coast-line of West Europe during early Post-Glacial time was outside our islands. Terrestrial conditions during this period were remarkably favourable for the emigration to our area of the fauna and flora which had not been exterminated by the preceding glacial period, or such portion of them as had survived its rigours, or were able to return. We picture the North Sea as a broad undulating plain once more, traversed by the Rhine with all our eastern rivers from the Thames northward as its tributaries; we picture the Irish Sea as another plain, more rugged, perhaps, in character, perhaps with a long central lake and an ex-current river, draining parts of Ireland, England, and Scotland. We picture the Bristol Channel as a verdant valley with precipitous sides, watered by a Severn flowing on to join the main river of the Irish Sea valley. Further, yet again, we must picture the English Channel as another broad plain watered by the Somme and the Seine, then one river, with tributary streams from the south of England, flowing south-west to fall into the Atlantic, whose coast-line then extended many miles to the west of Ouessant Island.

We appear to possess no evidence of the time that these geographical conditions endured. Sufficient time must have elapsed, however, for the climate to become so modified as to allow of the growth of a luxuriant flora and of the emigration of vast numbers of birds and animals from more southern regions to which their range had been contracted by extermination by the preceding glacial period. Submergence seems, however, to have been in progress, due either to terrestrial action or marine disturbance, and Ireland became separated from England before more than twenty-two of the forty species of British Mammalia, and four of the thirteen species of Reptilia and Amphibia had succeeded in establishing themselves in that area. So far as birds are any evidence the land must soon comparatively have been invaded by the sea, until the contour of 40 fathoms was reached. This would bring Ireland almost to its present contour and isolate it from England and from the Continent by a wide sea in the south, which was probably then the dominant line of Emigration of all species moving north with the changing climate. The 40-fathom contour, however, would leave Ireland joined to Scotland by a narrow isthmus—of no consequence as regards southern Emigration, as I hope to show in a future chapter—and submerge about half of the North Sea plain beneath the waves. The sea at this contour would encroach considerably upon the English Channel area, but still offer little obstacle to the northern or western Emigration of species by that route; whilst sufficient land would be left between Denmark and England to allow of Emigration and Migration on a vast scale, as the sequel will show. Proof of the rich mammalian fauna dwelling at this time in Western Europe is suggested by the great quantities of its remains which have been dredged from the bottom of the North Sea, especially from the Dogger Bank, from which latter locality bones, teeth, and antlers have been obtained, relics of the reindeer, Irish elk, stag, woolly rhinoceros, mammoth, hyæna, bear, wolf, etc., some of which must have roamed that ancient plain during prehistoric ages.

But the sea still continued to encroach upon the land, terrestrial submergence or marine upheaval steadily progressed. The North Sea crept southwards or spread east and west, and the Atlantic meantime encroached more and more up the English Channel, the Bristol Channel, and the Irish Sea, until the contour of 20 fathoms was reached, and the Dogger and other banks in the North Sea became islands, on some of which many large animals appear to have been imprisoned by the surrounding sea, and eventually exterminated in the final disappearance of this area beneath the waves. How long the contour remained at some 20 fathoms we have no means of ascertaining, but the final severance of the British Isles from continental land occurred at no very remote geological period, when the waters of the English Channel and the German Ocean encroached upon the narrow isthmus between France and England and finally met at the Straits of Dover. At first the newly-formed Strait was exceedingly narrow, perhaps less than a mile. For some time after the final severance of Great Britain from continental Europe nearly all the coast-line of England, part of Ireland, and part of Western Europe remained apparently at the 15-fathom contour, as is suggested by the phenomenon of submerged forests. At the 15-fathom contour the coast-line of the east of England from the Humber to the north of Norfolk would extend many miles further out to sea than is now the case; elsewhere it would only reach a few miles beyond existing limits, with the exception of a considerable distance outside the present coasts of Norfolk, Suffolk, and Essex. The coasts of the Netherlands and Belgium also probably extended further out to sea during this period. The significance of these facts will be made apparent in a later chapter.

Such, broadly speaking, are the principal physical changes which the British area has undergone from the close of the Pliocene epoch down to prehistoric time. In order to render the subject more complete and to enable the reader better to grasp these mighty changes in their natural sequence, I append a brief résumé of the events recorded, associating with them the various physical and climatal changes occurring contemporaneously in other parts of Europe, together with the probable effects of such changes on the co-existent faunas and floras. For this latter information especially I am chiefly indebted to the profound researches of Professor Geikie, one of the highest authorities on the subject.

We may begin our cursory sketch with the gradual passing away of the second glacial period—with the retreat of the ice-sheets from North-western Europe and the melting of the glaciers in the valleys of the Alps and other mountain chains of Central and Southern Europe. We remark the northern advance of plants and animals, their invasion of alpine regions, and their replacement by more temperate species from still more southern areas, emigrating northwards with the amelioration of climate. Dense forests gradually spread northwards and up the mountain sides, under the development of favourable climatic conditions; and a mixed and luxuriant flora slowly covered the long-desolated land. Once more the mammals emigrated northwards, many passing from Africa to Europe by land connections between Italy and Tunis, and Spain and Barbary; southern and temperate forms commingling and ranging from the shores of the Mediterranean up to the north of England, and even into Scotland; large carnivores frequenting the forests and bears the caves. "If laurels, fig-trees, and judas-trees grew side by side in Northern France with the sycamore and the ash, and in low-lying countries on the borders of the Mediterranean with pines, oaks, beeches, poplars, and elms, so also," writes Professor Geikie, "were elephants, rhinoceroses, and hippopotamuses, horses, oxen and deer, hares and rabbits, wolves, foxes, lions, and hyænas joint occupants of the same regions." The geographical conditions at this period were such that the coast-line of North-western Europe extended beyond the British Islands, reaching what we have seen was probably the 100-fathom contour, and in the Mediterranean region the land was distributed differently from what is now the case.

Once more a change of climate and of physical conditions are initiated by the coming on of another—the third—glacial period, and another great extermination of the fauna and flora is commenced. As the cold increased the southern forms vanished from northern areas, and as the rigours of the climate became more intensified, temperate and boreal forms became less northerly in their distribution, until ultimately the dwarf birch, the Polar willow, and Arctic mosses, saxifrages, and lichens inhabited the lowlands of Central and Southern Europe; the mammoth and the glutton, the reindeer, the marmot, and the musk sheep dwelt only upon the northern shores of the Mediterranean. During this era the sea is presumed to have invaded the plains of the English Channel and the German Ocean, whilst it is suggested that the low grounds of Prussia were submerged. The British area, with the exception of England from Yorkshire southwards, was for the most part covered with snow-fields and ice, and the surrounding seas were filled with glaciers; many of the mountain regions of Central Europe were glaciated; and even as far south as Gibraltar and Malta heavy snows and severe frosts prevailed. Again the severity of

15 20 & 40 FATHOM CONTOURS OF THE BRITISH AREA

Migration of British Birds Plate II.

CHAPMAN & HALL, LIMITED.

the climate abated, the ice-sheets of the north disappeared, the glaciers slowly shrank up the mountain valleys, and once more a general faunal and floral northern exodus commenced. The larger southern mammals—elephants, rhinoceroses, and hippopotamuses, hyænas, and the larger carnivores—however, never appear to have invaded Europe again, for no relics of their occurrence have yet been met with in the Post-Glacial deposits of that region. This absence of the large southern mammalia is very generally ascribed to the submergence of the land connections across the Mediterranean between Europe and Africa, but I think their disappearance from the European fauna of Post-Glacial times is due to other causes which will be discussed in a later chapter. We must omit from this period the great submergence of the British area, and in proceeding to discuss Post-Glacial phenomena, merely remark that with the passing away of this and the Baltic glacier phase of the Ice Age a very general elevation of previously submerged land took place, the 100-fathom contour being eventually re-established. This elevation would restore the plains of the North Sea, and perhaps connect Greenland with Europe by way of Iceland and the Faroes, not necessarily by a continuous coast-line, but by a considerable increase in the area of those islands. Following this elevation we have the series of submergences through Post-Glacial time which eventually culminated in the severance of the British area from continental land. I may here take the opportunity of remarking that the evidence furnished by the early Post-Glacial Scandinavian fauna is against the presumption of a direct land connection between Europe and

Greenland by way of Iceland and the Faroes. Such a land mass would effectually shut out the Gulf Stream from the shores of Norway, and render the climate of that area too rigorous for the luxuriant flora which palæontological evidence proves to have occupied the land after the Ice Age passed away. I would suggest a greater extension of land between Greenland and Europe in the form of islands; and that a greater land surface occupied this region during some portion of Post-Glacial time (after the third glacial period) is confirmed in a remarkable way by the migratory birds that follow this now submerged area, as we shall learn in a later chapter. Professor Geikie's views on the geography of this region during absolute Post-Glacial time (his first age of forests), in which he maps out Europe and Greenland united by an isthmus, seem therefore untenable.

Evidence for the greater extension of land between Greenland and Europe during early Post-Glacial time is furnished by the flora, and to a less marked extent by the fauna. So great an authority as Sir Joseph Hooker has shown the Scandinavian or North-west European character of the flora (through Iceland and the Faroes), and the paucity of Arctic American elements. So far as the mammals are an indication, we have the Arctic fox (remains of which have recently been found in the south of England) and the polar bear, found also in Iceland.[1] On the other hand, three other species are exclusively North American, although the fossil remains of one of them, *Ovibos* (the musk sheep), occur in the Post-Glacial gravels of Siberia, Germany, and France,

[1] There is, however, always the strong probability of these animals reaching Iceland on floating ice.

and in the brick earths of England—a fact in itself suggestive of a former land connection between Greenland, and Europe. Of the others, *Myodes torquatus* may have been introduced by man, and *Lepus glacialis* on icebergs None of the land birds, however, especially of South Greenland, show very striking Palæarctic affinities; and this to a great extent may be due to the present glacial conditions of the country, north of 65° on the east coast, combined with the wide water areas. The land birds indicate a more remote connection of South Greenland with Europe than with America, inasmuch as all the terrestrial species recorded by Hagerup (*Birds of Greenland*, 1891) as breeding in South Greenland are Nearctic forms of Palæarctic species, Palæarctic species that have invaded America from Eastern Asia, or species common to both regions. Iceland, as might be expected, shows more affinity with Europe in its avifauna, only one species belonging otherwise exclusively to the Nearctic region, viz. *Clangula islandica*. It also contains a Wren (*Troglodytes borealis*) found elsewhere only in the Faroes, and which is most nearly allied to the Norwegian *T. bergensis* and the St. Kildan *T. hirtensis*—a fact which strongly indicates a purely Palæarctic emigration. But the most interesting avian proof of a more continuous land area between Greenland and Europe is that furnished by the Wheatear (*Saxicola œnanthe*). The Wheatears breeding in Iceland and perhaps North Greenland are distinctly larger than the typical European form, and pass the British Islands regularly on passage during April and May, when the individuals of this species that spend the summer with us are already breeding. We are thus able actually to prove that this route is followed, the size of the birds and the date of their passage preventing any possibility of a wrong interpretation of the facts. The Wheatears or Chats (*Saxicola*) are exclusively an Old World group, confined to the Ethiopian and South Palæarctic regions, with the single exception of the Wheatear, so that there can be no doubt whatever as to the line of emigration followed by that species, or as to the more continuous land surface of its migration route whilst that route was being established.

The climate from the close of the third glacial period down to the dawn of historic time appears to have suffered considerable changes—alternations in fact of less and less strongly-marked genial and cold conditions, culminating in the fourth (and less extensive) glacial period, the epoch of the great Baltic glacier—as the eccentricity of the earth's orbit gradually diminished and approached that limit which characterizes the present era. There can be little doubt that the third glacial period was succeeded by a warm and genial climate, much more favourable to the growth of large vegetation than now, especially in the north, as is proved by the remains of forest trees in the peat-bogs of the now sterile Hebrides, Orkneys, Shetlands, and Faroes. Following this we appear to have a relapse to a colder period (the epoch of the Baltic glacier) than marks present time, followed by the submergence which was eventually to isolate our area from continental land, and also to separate Greenland from Europe by the wide water-ways that now exist. Local glaciers and snow-fields probably also appeared on the mountains of England, Wales, and Ireland, as they certainly did on those of Scotland. The climate during this period was not only cold but extremely humid; vast forest areas of oak, alder, and other deciduous trees were destroyed, and the ground converted into marshes and bogs, not only in the British Isles, but in Northern Germany and throughout North-western Europe. Eventually this cold, humid period passed away, and a second era of more favourable conditions supervened, during which a vigorous growth of forests (chiefly pines and birches) flourished. This era was marked by a gradual retreat of the sea. The subsequent destruction of many of these later forests and their burial in sphagnum peat mark a relapse to another period of cold, humid conditions, accompanied by a rise of sea-level and submergence of much forest-clad area in many maritime districts. To this latter period most of the submarine forests of the British coast-line are said to belong. With the passing away of this cold, wet period we approach the earliest ages of written history, the invasion of Britain by the Romans, the climatal conditions and arboreal surroundings which are characteristic of present time. The effects of these gradual yet enormous changes of climate upon the migration and distribution of birds in our area will never be known. That these changes must have influenced the fauna as well as the flora is unquestionable, but unfortunately these are details of our inquiry which will probably always remain beyond the limits of human knowledge or conjecture.

CHAPTER II.

RANGE BASE OR REFUGE AREAS.

Results of Climatic Change during Post-Tertiary Time—Condition of Europe during the Ice Age—Condition of the Mediterranean Area—North-west Africa during Pliocene and Post-Tertiary Time—The Saharan Region—Influence of the Sahara on the Distribution of Species—North-east Africa—Commingling of Palæarctic and Ethiopian Types—Homogeneity of the Fauna and Flora of Europe and North-west Africa—Pleistocene Land Connections between Africa and Europe—Effects of Glacial Epoch on the Euro-Asian Mammalia of the Miocene and Pliocene Eras—Commingling of Palæarctic and Ethiopian Types in the Sahara—Erroneous Views of Naturalists—The Canary Islands—The Cape Verd Islands and the Azores—Range Base or Refuge Area I.—Flora and Fauna—Range Base or Refuge Area II. Its Climate—Range Base or Refuge Area III.—Its Uses and Climate—Change of Climate in East Africa—Variations in Climate of Range Base or Refuge Areas Effects of Changed Climate on Avian Life—The Law of Dispersal—The Southern Exodus during Pleistocene Time a Myth "Arctic" Animals—The Cape Flora.

BEFORE proceeding further with our investigations respecting the dispersal of birds over the British area, it is of the utmost importance to our inquiry that we should endeavour to ascertain the character and amount of that physical and climatic change which took place contemporaneously in areas adjacent to our own. We have already seen that climatic variations during Post-Tertiary time have been enormous and severe. The inevitable result of such changes has been on the one hand to contract the range of species far to the south by extermination, on the other hand to encourage northern emigration, both movements being on a scale so enormous that we can only form a conception of their magnitude by a close and careful study of their results as manifested by the present distribution of existing species. What became of the numerous forms of animal and vegetable life dwelling in the areas suffering such glacial visitation? whence did they go during the devastation of their northern or alpine homes by the invading ice-sheets, snow-fields, and glaciers? Before we can hope to understand the philosophy of their distribution at the present time it becomes absolutely necessary for us to endeavour to trace out the areas these species occupied during the era of their extermination from glaciated areas, and from which they again set out as colonists upon the gradual return of more favourable climatal conditions.

Taking as our guide the present geographical distribution of the European fauna and flora we shall, I think, experience little difficulty not only in tracing out these areas of refuge, but in ascertaining to some extent the physical changes which those areas have undergone during late Pliocene and Post-Tertiary time. So far as Europe is concerned, the evidence I have been able to collect points to the existence of no less than three fairly well defined range bases or refuge areas. Europe, the western coast-line of which, following the contour of 100 fathoms, then extended outside the British area, at the climax of the Ice Age was to a great extent covered by one vast ice-sheet which descended to about lat. 50°

off the south coast of Ireland, and to about lat. 51½° or 52° across the south of England and Holland; thence southwards more or less irregularly to lat. 50° as far east as the northern Carpathians, whence it contracted northwards across Russia, here and there sending out furcations or spurs to about lat. 61° near the Urals. South of these limits were many local glaciers in Spain, in the Pyrenees, the Alps, the Carpathians, and the Caucasus. By far the most important of these vast ice-sheets were those in the Alps. For not only did they coalesce in the west with those on the mountains of Eastern France, but in the north probably with the Great Ice-sheet itself, through the glaciers on the mountains of Germany, Austria, and Bohemia. There can be no doubt whatever that during the climax of the glacial periods Europe was even more effectually divided into two portions by this vast central glacial system and its attendant snow-fields, than if a sea of equal breadth had stretched between them. There is evidence of two distinct glacial periods in the Black Forest, and, it is said, of three such glaciations in the Alps, separated by inter-glacial mild eras. During the maximum of glaciation or second of these glacial periods Central Europe, as we have already seen, was completely occupied by ice and snow-fields, the grand *Mer de glace* extending through the Black Forest into the valley of the Rhine and elsewhere, and coalescing with the glaciers of the Alps. During the third and the fourth glacial periods the northern ice-sheets appear not to have extended so far southwards, and did not deploy upon the low grounds beyond say lat. 52°, or coalesce with those of the Alps; so that the *Mer de glace* during the latter part of the

Glacial Epoch was broken, although it is probable that ice-sheets and snow-fields in Moravia and elsewhere helped to isolate Eastern from Western Europe, between the Alps and the Carpathians.

The condition of the Mediterranean and the distribution of land in that area during the Glacial Epoch are the next stage in our investigations. There is strong evidence to suggest that the Mediterranean Sea has endured throughout Pliocene and Post-Tertiary time; but, on the other hand, there is equally good ground for believing that the relative distribution of land and water in past ages has been very different from that which characterizes this region during present time. At a period no more remote than the penultimate interglacial era, and when the west European coast-line extended beyond the British Islands, it is suggested that Southern Spain was united to Barbary (Alboran Island probably then marking the western limit of the Mediterranean Sea); the Balearic Islands formed part of the Spanish mainland; Sardinia and Corsica were united and formed an isthmus which joined continental Europe in the Gulf of Genoa with North-west Africa; Italy, Sicily, and Malta were connected with each other and with the opposite coast of Tunis; the Adriatic and the waters surrounding the Grecian Archipelago were to a great extent dry land, extending southwards possibly to beyond Crete and Cyprus. The Mediterranean would then form three seas divided by the comparatively narrow necks of land formed by the Italian and Sardinian peninsulas being prolonged southwards into isthmuses joining Europe and Africa. What is now the northern portion of the latter continent was thus united to Europe by several important land masses which have now sunk beneath the waves, leaving Europe completely isolated from that region. We thus see that throughout Post-Tertiary time the Mediterranean has always been a very important barrier to the northern emigration and to the southern range contraction from Europe of vast numbers of plants and animals. This to my mind shows very clearly that the narrow strip of North Africa then attached to Europe was never occupied very dominantly by animals and plants whose European range was extensive northerly. It indicates that North Africa was then a land difficult of access in the only direction from which species could enter and colonize it by our Law of Dispersal, namely from the east or south-east. As we know, the great Sahara Sea between Tunis and Egypt must have barred all emigration to North Africa so long as it continued, and when this barrier was removed, an extension of range southwards for all northern forms would have been necessary to colonize this area. Hence the total absence of all boreal forms from North Africa.

From the physical characteristics of the Mediterranean area during the Glacial Epoch we now pass on to the opposite continent of Africa, and endeavour, with the evidence furnished by the present geographical distribution of European birds, to trace out its physical conditions during Pliocene and Post-Tertiary time. The present distribution of species seems very forcibly to suggest that Africa during those remote ages was divided into two unequal portions; or, rather, that the whole of North Africa, above the Sahara, including the Canaries, formed part of continental Europe. At that period the Sahara was either a great eastern inlet of the Atlantic, a relic of that vast primeval ocean which we know swept from the Atlantic to the Bay of Bengal during the Eocene and Miocene ages; or a vast sterile waste even more barren than at the present time, owing to the more recent retreat of the sea. Whether the sea actually invaded this area or not during Pleistocene time is, however, of little consequence: the fact remains that a vast barrier, either of sea or barren desert, effectually prevented all northern progress in this direction of plants and animals from Africa, and that the southern range of all western and temperate species never extended below this area. The vast influence of the Sahara on the distribution of species continues to be exerted down to the present time, and the strictly Ethiopian fauna and flora have almost entirely failed to establish any appreciable element amongst the Palæarctic types that still continue to occupy Africa north of that sterile barrier. In the north-east of Africa, however, we find a very different state of things, Ethiopian types ranging right up the Nile valley to the shores of the Mediterranean. Owing to the far easier conditions of dispersal we also find much commingling of Palæarctic (more correctly described, however, as of Ethiopian origin) and Ethiopian types in this area, the former penetrating the Ethiopian region to its southernmost limits, and from which region they originally emigrated northwards, and the latter encroaching upon the Palæarctic region to an appreciable extent. The Ethiopian element persists throughout Egypt even to the delta of the Nile; and the range of between thirty and forty species of birds characteristic of that region extends to Palestine, amongst the most notable forms being species of *Cossypha* and *Dromolæa*. We must not view these forms as remnants left behind in the north when the great and entirely legendary exodus into Africa was in progress, but either as Inter-hemisphere (*conf.* p. 60) species that have spread from an Antarctic, or Southern Hemisphere, centre of dispersal northwards across the entire Ethiopian region, or from an equatorial base. In Palestine, it may be remarked, we also find a commingling of types peculiar to the Oriental and the Ethiopian regions in such genera as *Nectarinia* and *Pycnonotus*. Such facts demonstrate very clearly that where the land surface is continuous and sufficiently fertile to support life, a commingling of types peculiar to different faunal regions invariably occurs, but in the Northern Hemisphere the invasion is never of a purely Southern character, the encroaching species conforming entirely to the Law that forbids a Southern extension of area (*conf.* p. 60), and also confirm the view previously expressed that the Sahara, either as an inland sea or a sterile desert, checked all such mixture of Palæarctic and Ethiopian types in the west of Africa, the division between the two faunas, so far as we know in that district, being a very abrupt one, whilst in the east of Africa, where no such barrier exists, the faunas of the two regions blend in a very suggestive manner. It is a most remarkable fact that the fauna and flora of Europe and North-west Africa have retained their homogeneity although separated by one of the deepest, widest, and most ancient seas in the world—phenomena that suggest a long-enduring and recent land connection, which we know existed at no remote epoch, as the palæontological remains on certain islands and the presence of submerged banks in their vicinity between the now disunited continental areas, undoubtedly prove. No such land connections ever appear to have bridged the primeval Sahara sea, and consequently the links between the typical African and European faunas and floras are insignificant and exclusively of recent appearance.

It now becomes necessary to revert for a time to the land connections between Africa and Europe during the Pleistocene Period. There can be little doubt that an extension of land surface across the Mediterranean was taken advantage of by birds, which seem always adverse to extend their range beyond water areas. These land passages across the Mediterranean have been chiefly insisted upon by geologists to account for the presence of the large mammalia in Europe during the Pleistocene Period, and for the presence in the North-west African flora of a decided European element: the Mollusca of the Sahara have also, I believe, exclusively Palæarctic facies. But, as I hope ultimately to prove, these land connections between South Europe and North-west Africa were incapable of preserving from extinction the large pachyderms and carnivores so characteristic of Pleistocene time. A glance at the map will show the reason for this. The area to the south of the Mediterranean from which these ancient land passages led—say between Spain and Italy respectively—must have been very restricted, confined at most within as narrow limits as at the present time, and bounded by sterile deserts or wide seas marking the southern limits of all the West European fauna and flora. Not only was the area an exceedingly limited one; it was also singularly ill-adapted to the requirements of such hordes of mighty beasts. North-west Africa possesses no large rivers in which the rhinoceros and the hippopotamus could find congenial haunts; no vast forests and wild fertile areas suitable for the elephant, the sabre-toothed tiger, and the mammoth. That many of these larger mammals inhabited North-west Africa during the Pleistocene Period, when the area was continuous with northern continental land, we have abundant proof in the remains which have been discovered in the caves of Algeria of *Equus*, *Bos*, *Antilope*, *Hippopotamus*, *Rhinoceros*, *Ursus*, *Canus*, and *Hyæna*; but how few of these have left descendants, and succeeded in preserving their species or their types in that area when it became isolated!

It is generally presumed that these large mammals retreated south in Africa to the haunts in which they live now, and which were then occupied by many of the same species, coming north again with the return of milder climatal conditions. But there can be little doubt that many of these animals dwelt in the vast Euro-Asiatic continent which then included North-west Africa, over which they were widely dispersed as a dominant fauna, the range of at least some species extending northwards from tropical Africa, as that area was united to the northern continent by the upheaval of part of the bed (from Egypt and Abyssinia to the delta of the Ganges) of that primeval ocean which then extended from the Atlantic to the Bay of Bengal, this event taking place in late Miocene or early Pliocene time. As we have already seen, the western portion of that ancient ocean perhaps endured across what is now the Sahara throughout the Glacial Epoch. As a proof of the existence of this Sahara sea at no very remote geological era, I may remark that the remains of species of marine shells, still existing in the Mediterranean, are found scattered over the desert, both in the valleys and up to elevations of 900 feet; whilst at least one species of fish has been found in the salt lakes of that region, south of Algeria, which is also an inhabitant of the Gulf of Guinea, nearly 2000 miles to the south! The large quantities of salt so characteristic of the whole region, as I can testify from experience, may also be safely regarded as direct proof of the recent existence of a vast salt-water area.

I hope I have succeeded in making it tolerably plain to the reader that the great exit from tropical and South Africa for species increasing their northern range was in the north-east, down the valley of the Nile; and it is by this route, and this route alone, that we trace the ancient range contraction of such species that had penetrated far into the Palæarctic region. Even this area is now, as it undoubtedly was during Pleistocene time, not very favourable to the emigration northwards of Ethiopian mammals and plants. There can be little doubt that the Glacial Epoch was the cause of the extermination of all or nearly all the great, varied, and dominant mammalian fauna that roamed in that vast Euro-Asiatic continent during Miocene and Pliocene ages; and the utter absence of the remains of the great pachyderms and carnivores from Post-Glacial deposits in Europe (with the one doubtful exception of the mammoth) must not be ascribed to the submergence of the land passages between North-west Africa and South Europe during the last glacial period, but as an eloquent and uncontrovertible testimony to their extermination through adverse conditions of life, due not only to severities of climate but to narrower areas of distribution. As Professor Geikie himself remarks: "We have seen that a number of characteristic Pleistocene animals had made their appearance in England at the close of the Pliocene Period, or shortly before the advent of the earliest recognized glacial epoch of Pleistocene times. They were associated with several Pliocene forms, such as *Hippopotamus amphibius*, *Elephas meridionalis*, *Machairodus latidens*, *Rhinoceros etruscus*, *R. megarhinus*, *Ursus arvernensis*, *Cervus dicranios*, and *C. polignacus*. Of these, one, the hippopotamus, is still living, while others do not appear to have survived in North-western Europe the first glacial epoch. The southern elephant and the megarhine rhinoceros, however, struggled on into inter-glacial times, when the former occupied the valley of the Rhone, Central France, and Northern Italy, and the latter ranged from Southern Europe into England. The sabre-toothed tiger also would seem to have persisted well on into the Pleistocene Period. Of the other animals that come into view for the first time in the pre-glacial deposits of Cromer, many appear and re-appear in successive inter-glacial deposits; but we note as we advance toward the later stages of the Pleistocene that some of them become rare, while others vanish altogether. The recurrent glacial epochs seem to have told severely upon many of the herbivorous animals. The only two pachyderms that have survived are the hippopotamus, already mentioned, and the African elephant. During each successive glacial epoch those species which could only exist under a mild climate would be forced to the extreme south of Europe, where, confined within ever-narrowing limits, they would gradually die out. Only the more robust types, such as stag, megaceros, urus, bison, horse, mammoth, woolly rhinoceros—species capable of enduring some severity of cold—would live on. The carnivores, however, might be expected to thrive wherever their food supply was sufficiently abundant, they would prey alike on the denizens of the southern regions and the occupants of less temperate latitudes. Most of them, therefore, were enabled to endure all the climatic vicissitudes of the glacial period, and many are recognized as still living species." As will be seen, but one of these pachyderms has managed to survive the Glacial Epoch, and that is now an inhabitant of tropical Africa. To the present writer there seems no reasonable doubt that this surviving species, the hippopotamus, is the descendant of individuals that entered Africa in late Miocene or early Pliocene ages, when that region was invaded by so many of the higher species of mammalia either from the east or west, and that the Ethiopian portion of this species never had any experience whatever of the Glacial Epoch. The African elephant is said to be specifically distinct from the European elephants of Pleistocene time (perhaps northern and southern forms), but its remains, however, have been found in North-west Africa, individuals which were isolated there during the glacial climate of Europe and exterminated by adverse conditions. Had conditions been favourable for emigration to Africa of the various large forms of Pliocene mammalia, there can be no reasonable doubt that many species would have availed themselves of such a retreat to the tropics and survived amongst such congenial surroundings down to the present day, yet, so far as I can ascertain, there is not a single trace throughout the Ethiopian region of any such Pleistocene exodus, for the Law of Southern Dispersal forbids it. The evidence also points to the conclusion that the same Law forbade the retreat of these Euro-Asiatic mammals into Africa by way of the Nile. The presence in tropical Africa of the hippopotamus, the sole surviving form, undoubtedly proves that the range of this species during the Glacial Epoch was Ethiopian. Precisely the same remarks apply to birds. We have abundant evidence of the former existence in Europe during Tertiary time of many Oriental and Ethiopian types, such as *Collocalia*, *Trogon*, *Psittacus*, *Necrornis*, *Limnatornis*, *Centropus*, etc.,[1]—all species that had spread northwards and eastwards from southern or equatorial bases and become sedentary in the warm equable climate of Euro-Asia until exterminated by the advent of adverse conditions of life, primarily due to a change of climate; and that now continue to be solely represented by types that have preserved their identity

[1] Every species that did not BREED south of glacial limits must have perished; numbers of other species that were resident in many parts of Euro-Asia must also have been exterminated by the adverse change in the climate. Among these were such types as are here instanced. If the advent of a glacial climate could initiate a migratory southern movement or an extension of permanent habitat in the same direction, then numbers of such types would have survived and left traces of their southern exodus behind them; but adverse climatic conditions can never lead to increase or extension of winter or non-breeding area, and extermination is the inevitable result, unless modification may in some cases preserve them from utter obliteration.

because they were never subjected to such vast exterminating conditions as have entirely removed every one of these their northern descendants. If the range base was beyond the limits of exterminating influences (either connected by migration or continuous breeding area) the species or the type survived; if entirely within such fatal limits, just as surely did such types or species vanish for ever; for the Law of Dispersal (*conf.* p. 60) in the Northern Hemisphere forbids all southern emigration either to increase breeding area, or to escape adverse climatic change, or unfavourable conditions of life. One of the most graphic examples of the inability of a fauna to escape from utter extermination is afforded at the present time in New Zealand. Let the reader peruse Mr. W. H. Smith's sad account of the extermination of birds in that country (*Ibis*, 1893, pp. 509—521), and he will be speedily convinced of the truth of the axiom that species never "retreat" from unfavourable conditions only by extermination. In the case of a wide-ranging dominant species, only such portions of the species are affected as are subject to the adverse conditions; but if the species occupies a narrow area nothing can save it from utter extermination if the adverse conditions continue.

Of the rich and magnificent fauna of Euro-Asia what an utterly insignificant remnant has been preserved to us, how devastating has been the result of the Glacial Epoch upon these, some of the highest as well as the more archaic forms of life! The present distribution in Africa of many of the larger carnivores (such as the lion, panther, cheetah, jackal, and hyæna) may, however, be slightly due to a Pleistocene range contraction from

Europe during the last glacial period. After the range of these animals had been contracted into Africa by extermination down the land passages, if you will, which existed between that continent and Europe, we can readily understand how a portion of these species might have been left isolated in North-west Africa after the submergence of such passages, and consequently never re-appeared by Post-Glacial Emigration in Europe with the return of milder climatal conditions. Here, subjected to comparatively easy conditions of life, for abundance of food could be obtained, many of them were doubtless able to subsist even to the borders of the Southern Sea or arid desert; and as this area eventually became studded with oases and scattered vegetation, the descendants of the individuals of many of the same species which entered Africa in Miocene or early Pliocene ages gradually spread northwards or westwards, so that the distribution of the species, once discontinuous, eventually became uninterrupted as at present. The Sahara has unquestionably been peopled from the east and south. Emigration commenced from both points as soon as life could be supported there; from the east by species in search of an extension of range, and from the south by species similarly extending their areas of dispersal northwards in summer, or during the period of reproduction. The presence in this area, however, of endemic forms of these southern and eastern animals, seems to suggest that the colonization of the old Eocene sea-bed in the south was not very extensive, and took place under more or less unfavourable conditions. Even where mammalian and reptilian affinities are not with European species they are with Asiatic forms and not with Ethiopian ones, another proof of the complete isolation of the old African fauna. It is impossible to say in what portion of the Saharan region the Palæarctic and Ethiopian faunas intergrade, or whether they do so at all to any appreciable extent, until we possess more complete information respecting the animal life of this profoundly interesting area. Any one reading Mr. Sharpe's recent paper on the distribution of birds (*Natural Science*, 1893, p. 105) would very naturally infer that the Ethiopian element penetrates into Morocco. That naturalist remarks: "It is a somewhat important fact that a truly Sudanese form like *Melierax polyzonus* has been found in Mogador and Southern Morocco." In the first place it is an error to describe this bird—one of the chanting Hawks—as strictly Sudanese, seeing that it is distributed over the whole of Africa, except the Desert, the north-coast countries, and Egypt. It is erroneous and misleading thus to infer that the Sudanese element extends to Mogador because this bird has only accidentally strayed into Morocco, as I was assured by the late Mr. Gurney, then our highest authority on the *Raptores*. As well say with equal propriety that the Nearctic avifauna coalesces with the Palæarctic avifauna in the British Islands because an example say of the Yellow-billed Cuckoo or the American Bittern has been captured in them! Dr. Wallace, in the latest edition of his *Island Life* (pp. 396, 397), gives a long table of "Land birds common to Great Britain and Japan," and he is careful to inform us in a footnote that "accidental stragglers are not reckoned as British birds." Yet, incredulous as it may appear, he actually includes such species as the Nutcracker, the

Rustic Bunting, and the Eagle and Scops' Owls! Dr. Wallace states that this list "is very interesting when we consider that these countries are separated by the whole extent of the European and Asiatic continents, or by almost exactly one-fourth of the circumference of the globe." Out of the fifty-three species enumerated eleven have no right whatever to be classed as British at all, but are abnormal or nomadic migrants from the far East, from the South, or North; whilst in the other instances the area of distribution between Britain and Japan is continuous, or intermediate sub-specific forms, due chiefly to climatic variation, constitute an unbroken link between species common to both countries! Such statements are terribly misleading to the student, shake his confidence in work in which he is not sufficiently expert or has not the requisite special knowledge to test its accuracy, and illustrate very forcibly how absolutely necessary it is that naturalists should thoroughly understand not only the rudiments but the higher philosophy of the Geographical Distribution of Life, before they attempt to theorize upon it or endeavour to demonstrate it. *Ne sutor ultra crepidam* just as aptly applies to scientists in general, and to Royal Institution lecturers in particular, as to shoe-makers or any other craftsmen.

It now becomes necessary to glance briefly at the Canary Islands and the Cape Verds. The former group is situated off the coast of West Africa between the 27th and 29th parallels of latitude, and the nearest of the islands are not much more than fifty miles from the continent. They are separated from Africa by a channel more than 830 fathoms in depth, are mountainous, and are entirely volcanic. Dr. Wallace is of the opinion that they have never been connected with the adjoining continent, and bases his belief chiefly on the absence of all indigenous land mammalia and reptilia. I am, however, inclined to believe that these islands have experienced a continental period in remote ages, and that they are but the isolated terminal portion of the great Atlas range. The vast elevation of the ocean bed required (some 5000 feet) to unite them now with continental Africa is not so much as Dr. Wallace himself shows necessary to join the Philippines with continental Asia (some 6000 feet), and but little more than what must have occurred to connect certain land surfaces in the eastern Mediterranean, of which we have direct palæontological proof of union during Pleistocene time. Further, it is a well-known fact that local marine depressions of vast depths are by no means uncommon features of regions subject to volcanic action. The absence of terrestrial mammals and reptiles does not appear such an insurmountable obstacle when we remember that the area during a past continental period was a very isolated one, and that the adjoining parts of Africa even during Pleistocene time were too arid, desolate, and barren to support a very large mammalian or reptilian fauna, and the probabilities are that it did not contain any such animals at all in areas contingent to the Canary group. I would suggest a submergence (perhaps of a sudden, rather than of a gradual character) some time during the Pleistocene Period, after such desert-loving species as the Courser, Houbara Bustard, and Sand Grouse (dominant desert forms) had gradually spread westwards with the slow emigration of life over the recently elevated Sahara. The avifauna of the islands is with one exception entirely of Palæarctic derivation, and we have direct proof that they were inhabited by Palæarctic species throughout Pleistocene time, when the breeding range of birds was contracted by extermination down to North-west Africa by the devastating ice-sheets which descended upon northern and temperate Europe. The only Ethiopian element in the Canarian avifauna is the Black Oystercatcher (*Hæmatopus capensis*). There can be no doubt that this species is a very recent emigrant to the islands—a bird gradually extending its range northwards along the now continuous coast-line of Africa; for we have strong evidence to suggest that when the breeding range of birds of this genus was contracted south by the Glacial Epoch it had no base down the African coast-line beyond the Canaries, the ancient ocean defining the range limits in that direction (see Map), but had spread westwards across North Africa from an Ethiopian or Asiatic base. The Black Oystercatchers appear to have been differentiated in and dispersed from a southern hemisphere base. To the present time the Canary Islands are the winter resort of numbers of European species; they contain many insular forms or representative species of Temperate European or North African forms, and a number of species range from these islands eastwards across the whole of North Africa—facts which seem to suggest a colonizing movement during a remote period of continuous land surface, rather than a fortuitous western emigration (of at least 50 miles across the sea), especially when we bear in mind the reluctance of birds to cross a wide water area in extending their range from centres of dispersal.

The Cape Verd Islands are a much more debatable area. They are probably oceanic islands of great antiquity, separated from continental Africa and from themselves by a sea some 7000 feet in depth. They are also extremely isolated, lying between 300 and 400 miles from the coast of Senegambia. Unfortunately, we possess few reliable particulars respecting the fauna of these islands. There can be little doubt that they have been entirely populated by fortuitous emigration from the Palæarctic and Ethiopian regions. There is certainly a strongly-marked Palæarctic element in the avifauna, combined with several Ethiopian forms, some of the species very clearly indicating the line of emigration followed. The islands are situated off the African coast, a little to the south of the Sahara, and probably in a direct or nearly direct line with the ancient southern shore of that sea which once covered the Desert. Some of the birds appear to have emigrated along the southern coasts of this ancient sea, or, in more recent times, along the borders of the Desert from more southern areas in Africa. It is true one or two species of birds are found in the latter islands as well as in the Cape Verds, but there is nothing to suggest a southern emigration; whilst on the other hand we have absolute proof of a north-western emigration from the south. Thus the Blackcap Warbler (*Sylvia atricapilla*) is only known to resort to the Canaries in autumn, when large numbers visit the islands and swell the resident population of this species. Now, did the birds pass further south by this route they would be seen in spring coming north again, but no such migration has been remarked. This Warbler passes the Nile valley and certainly goes as far south in East

Africa as Abyssinia. There can be no reasonable doubt that one of its fly-lines passes south of the Desert to Senegambia in West Africa, where the bird is also found in winter, and from which an extension of range commenced to the Cape Verds. Again, the Spectacled Warbler (*S. conspicillata*) has probably reached the islands by a similar route, seeing that it is a resident in the Canaries, but winters far into the Desert. There are also several typical desert species found upon the Cape Verds, such as *Certhilauda desertorum* and *Ammomanes cinctura*, Palæarctic or Oriental in affinities, and which have reached the islands by fortuitous emigration from those regions. As regards the Ethiopian elements we have a Kingfisher (*Halcyon erythrorhyncha*), very nearly allied to the *H. semicærulea* of Arabia and North-east Africa, further indicating an emigration along the southern borders of the Sahara; whilst *Accipiter melanoleucus* and *Pyrrhulauda nigriceps* seem to suggest an emigration to them from an equatorial base, seeing that the former species is now found throughout Africa south of the tropic of Cancer, and the latter is said even to occur as far north as the Canaries. We shall have occasion to allude to this migration route across Africa from the north-east to the west in a future chapter. I may here state that it is exceedingly improbable that *Corvus corone* and *Milvus ictinus* ever normally visit the Cape Verds, although both have been included in their avifauna.

The Cape Verds (and we may also say the Azores), therefore, although decidedly Palæarctic in the facies of their fauna (including the Coleoptera and certain genera of land shells), can scarcely be regarded as part of any

RANGE BASE OR REFUGE AREA I. Migration of British Birds. Plate III

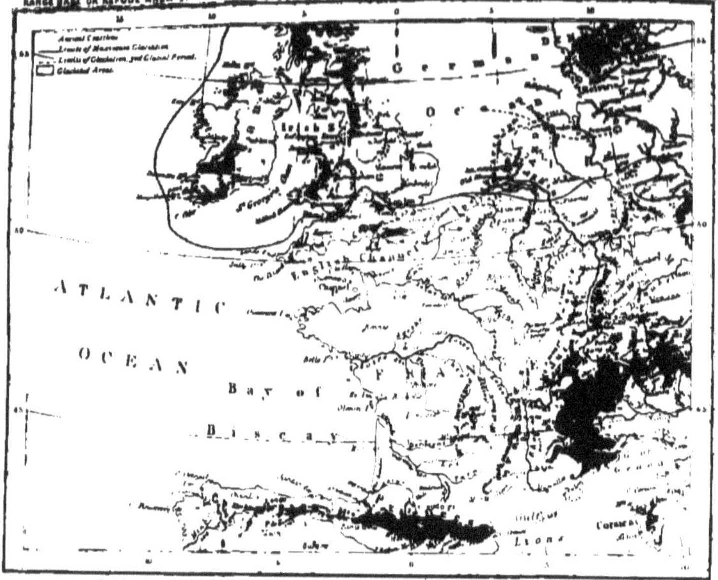

CHAPMAN & HALL LIMITED

glacial refuge area or range base, but owe their present forms of animal life purely to fortuitous emigration. We do not yet possess sufficient knowledge of the Cape Verd Islands to say whether any Palæarctic birds regularly winter in them; if such be the case, I should expect, judging from the evidence at present available, that they were reached by a south-westerly migration route from South-western Asia and North-eastern Africa, rather than by a direct southerly route from Europe and North-western Africa. The sea passage is much too wide to expect any regular migration to this purely oceanic area; although if the species remain constant in that area the fact would seem to imply pretty frequent visits, intercrossing preventing any tendency to segregation due to isolation. It is probable that island forms do exist in the Cape Verds, especially when we know they are so prevalent in the Canaries, a far less isolated area. I have dwelt at some length upon the probable condition of North Africa during the Pleistocene Period, because it is of great importance to our inquiry, as we shall learn eventually when we come to enter into details respecting the distribution of British birds in that area.

As the preceding evidence suggests, the three great RANGE BASES or Refuge Areas of British birds during the Ice Age may be defined as follows:

RANGE BASE OR REFUGE AREA I.: This area included all the now submerged land in the English Channel to about W. long. 10°, and from thence southwards along the western coast of France to Biarritz. The land of this area which still endures consists of England, say south of the Thames and the Bristol Channel,

Belgium, and Germany, about as far as E. long. 10°, together with that greater portion of France which remained free from permanent glaciers and snow-fields. Thanks to the researches of geologists and palæontologists we can form a pretty good general idea of the state of this area during the Glacial Epoch. At the climax of the Ice Age, and probably for ages after that event, there can be little or no doubt whatever that bird-life was utterly exterminated from all parts of Western Europe, including the British Islands, say north of lat. 52°. The present area may be described as the head-quarters of those now high Arctic species that in those remote days contrived to live and thrive and perhaps become modified into "boreal" forms amidst the rigours of the glacial ages, practically on the margins of the snow-fields and ice-sheets. The flora of this region was very similar to that which characterizes Scandinavia and the high Alps at the present time, perhaps the most characteristic Arctic plants of Southern England then being equally modified mosses, lichens, saxifrages, and the dwarf birches, and willows of various species, as were, for instance, ascertained by the late Mr. Pengelly and Professor Heer, and again by Mr. Nathorst, to occur in the clays of Devonshire. The reindeer and the elk, the lemming, the glutton, the Arctic fox and the pika—all now thoroughly Arctic species—were dwellers in this area, as their remains have so eloquently demonstrated, and with a range base in that area which must have endured from Pre-Glacial times, or a dispersal thereto from more eastern and southern non-glaciated areas during the continuance of glacial conditions elsewhere (*conf.* Map, p. 47).

RANGE BASE OR REFUGE AREA II.: This area included all the now submerged land in the Mediterranean west of say E. long. 20°, and extended as far south as the northern coast-line of that ancient sea which once occupied the Sahara. It also included the peninsula of which the Canary Islands and Madeira now form the few lingering relics, and the additional Atlantic coast area which would be obtained by making the contour of 100 fathoms represent the possible coast-line of this period. The land of this area which now endures consists of the Spanish peninsula, Alboran Island, the Balearic Islands, Corsica, Sardinia, Italy, Sicily, Malta, a portion of Tripoli, the whole of Tunis, Algeria, and Morocco, together with the Canary Islands. This area is by far the most important one, so far as British birds are concerned. It formed the chief, nay probably the only range base of nearly all the species that now breed in our islands, as well as of most of the avian life that had been exterminated by the glaciers and climatic changes in all regions lying directly to the north of it. The climate of this region in the southern portions was probably as genial as now, even at the climax of the glacial era; whilst in the northern portions, although at that period more rigorous than now, it was sufficiently genial to admit of the constant existence of temperate plants and animals within its limits. The summers in the northern portions of this area were probably always as favourable to bird-life as now, nay even more so, if at times they were somewhat shorter; whilst the winters of the southern portions were always marked by those genial conditions which are so imperative to the needs of such vast numbers of species (*conf.* Map, p. 71).

RANGE BASE OR REFUGE AREA III.: This area included all the now submerged land in the Mediterranean east of say E. long. 20°, and Europe south of about lat. 47°, and east of the Adriatic. At this period probably the whole of what is now the Ægean Sea was dry land, the coast-line extending along the western shores of Greece and continuing unbroken or nearly so along Crete to Cyprus, and thence down the coast of Syria, which possibly extended seawards much further than is now the case. South of the Caucasus, Georgia, Armenia, Asia Minor, Syria, Northern and Western Arabia, Persia, and Afghanistan, must be included as part of this refuge area; as also must the entire continent of Africa as it then existed, its northern coast-line reaching from the Atlantic along the southern limits of the ancient Saharan sea. The chief physical changes in this area have been on the one hand the vast series of submergences which have produced the Grecian Archipelago, and isolated Crete and Cyprus; and on the other hand, the retreat of the ocean from the Sahara with the accompanying increase of the land surface of continental Africa in the north. So far as concerns British species this area is the least important of the three. We have the most positive evidence to show that the range base of purely West European birds did not extend to that region during the Ice Age to any important extent. It is, however, absolutely necessary for us to recognize such an area, for it formed the range base of certain species which, in ultimately extending their area to the British Islands during Post-Glacial time, present some very curious instances of dispersal, and forcibly illustrate the eccentricities which sometimes characterize the migration routes of birds. The northern portions of this area probably differed little, if at all, in their climate from portions of the preceding refuge area within the same parallels of latitude; the climate of the southern portions we may reasonably infer was not influenced to any great extent by the Glacial Epoch, although mountainous regions have unquestionably suffered some local modification in this respect. Proof of geologically recent change of climate in East Africa was obtained by Mr. J. W. Gregory in his expedition to Mount Kenya. Mounts Kilima Njaro and Elgon, the mountains of Abyssinia and the Cameroons, I believe I am correct in stating, furnish additional proofs of local climatic change, due to more intense glaciation in remote ages. There is also much evidence to suggest that North-east Africa at no geologically remote epoch was much more wooded than is now the case—conditions that would favour the western emigration of arboreal forms (*conf.* Map, p. 95).

Of course it must be clearly understood that the climate of these refuge areas, most especially in the first, underwent considerable variation during the course of the Glacial Epoch, which, as we know, was made up of a series of alternate mild and cold periods. Mild climatal conditions were succeeded by eras when the cold was more intense, and these in their turn gave way to warmer ones; at one time the summers must have been short and hot, the winters long and intensely cold; at another, longer and colder summers prevailed, and the rigours of winter were less severe, or the seasons were more blended; whilst yet again the pluvial conditions of the warm season were much intensified during certain epochs. Vast plain and valley floods were the summer characteristic of some of the glacial ages, and enormous tracts of low-lying country were inundated, as for instance in parts of Middle Europe and Southern Russia, in the midst of which birds must then, as we know they do now, have found a perfect paradise, teeming with food. Similar floods, but on nothing like so gigantic a scale, caused by the rapid melting of vast quantities of snow and ice, continue to be one of the most characteristic features of the continental Arctic regions, especially in the river valleys, where the break-up of the streams in the scarcely-perceptible northern spring is the grandest natural phenomenon of each recurring year. The melting of the snow from the Alps, causing vast areas to be annually flooded, is another present day instance. The higher latitudes of course, then as now, were subject to the greatest cold. The range of birds throughout the varying phases of the Ice Age must therefore have expanded and contracted north and south as the climate alternately favoured or enforced one movement or the other—oscillated as it were between northern and southern areas with each mild or cold Glacial Period. It will probably ever remain a hopeless task to attempt to portray those avian movements, or to indicate the probable species that made them. Our only chance of success seems to rest with the close of the glacial era; with the last cold periods of the Ice Age which contracted (by extermination or restriction of northern breeding areas) avian distribution southwards, and with its ultimate expansion across those desolated northern areas, as the cold passed away, and was succeeded by milder climatal conditions. Broadly speaking, the last range contraction south and its expansion north are a facsimile of those phenomena that preceded them.

I may say in conclusion that so far as I have been able to ascertain, many of the views expressed in the present chapter are entirely new. I maintain that the immediate Pre-Glacial dominant fauna (including Aves) and flora of temperate Euro-Asia were never very closely associated with Africa, or with Australia; probably nothing nearly so much as were the immediate Pre-Glacial fauna and flora of North America with those of South America. Professor Geikie writes: "The European flora of to-day differs so much more from that of Pliocene and Miocene times, than the flora of North America does from the plant life of those periods. In the latter continent there existed a continuous passage to the south across which plants and animals alike could make good their retreat." Now to my mind this very conclusively proves that the line of Antarctic connection is most dominant along the American continents, and that the floras were more continuous than in the Old World. Unfortunately, Professor Geikie (to whom this Law forbidding southern dispersal or emigration was unknown) makes this continuous land mass a proof of Southern Emigration rather than a contraction of range. The Law that forbids southern dispersal in the Northern Hemisphere prevented a southern emigration of all the purely Euro-Asian species, and consequently the Glacial Epoch nearly exterminated that fauna and flora. Hence the abundance of its palæontological remains bears eloquent testimony to the helplessness of such animals to escape from their glacial doom. A few relics only survive whose range then extended right across that region from north to south. No purely Polar type could have survived this or any other glacial epoch—not even that one famous Præ-Pliocene ancestor of the Charadriidæ. Species, however, that were of South Palæarctic, Ethiopian, Oriental, Australian, Antarctic, South Nearctic, or Neotropical origin survived the vast exterminating influences of the Ice Age, their northern range simply contracting by extermination more or less upon the southern base of those species, or what are now their ancestral forms. Hence the reason why the early Europern flora was a portion of that which now exists only in the tropical and sub-tropical lands of the Eastern Hemisphere. There could have been, therefore, no southern exodus of all living things, as scientists so unanimously assert there was; no trace whatever of that southern emigration during Pleistocene time exists, and the only species that succeeded in maintaining themselves throughout that era were those whose range was either cosmopolitan throughout the lands whose northern portions were affected, or species whose emigrations had extended from a far southern base, and were then continuous with that base, either by extension of uninterrupted breeding area, by subspecies or representative races, or by migration from and to a southern base, the winter range coalescing with that of summer. One apparent difficulty is presented in the fact of high Arctic animals occurring only in the south during glacial eras, which looks like a purely southern emigration. We have no absolute proof that the southern bases were ever entirely deserted by these what we now class as boreal forms, and I would suggest that these species had endured in the south during milder intervals of climate, by portions of them ascending to great altitudes on mountains as migrants, which again descended to the lowlands as residents as the climate again became glacial. Southern mountains were in fact the grand preservers of the species in the south during warm intervals of climate, just as we find to be the case to-day. It is by this means probably many types or species have succeeded in maintaining their existence during the Glacial Epoch; many instances might be given of boreal birds breeding at high elevations on southern mountain ranges, and wintering on the lowlands, or migrating much further south to an original base. The migratory tendencies of all Arctic animals is significant. These animals are Arctic in the sense of having become Arctic since the Glacial Epoch passed away. There is no evidence whatever to suggest that they were confined exclusively to the Polar latitudes during Pre-Glacial time. Arctic animals are species that have spread north from southern bases during Post-Glacial time, not species that have been driven from a previous Polar haunt by glacial climates. They are species that have slowly adapted themselves to Arctic conditions, as their range has been extended north by Post-Glacial emigration into Polar latitudes. Note the various Arctic forms or representative races of more southern species. Arctic climates would produce Arctic forms. This by no means implies that Arctic species emigrated south, but that a glacial epoch exterminated them, and that the effect of a much lower latitudinal Arctic climate prevailing during glacial epochs had its usual effect on species, or such portion of species that came within its influence, and were able by modification to withstand its rigours. Speaking generally, Arctic climates do not affect *structure* so much as *dermal covering* and *colour*; so that fossilized remains of these reputed boreal types are by no means a complete record of their characteristics.[1] It may therefore be safely prophesied that all peculiar boreal forms now living exclusively in the Arctic regions would perish in a glacial epoch, if that glacial climate included the southern limits of their present range. The reindeer would perish as surely then as the Irish elk has utterly vanished in the past. This Law of Dispersal also explains why none of the Euro-Asian deer and bears have penetrated into the Ethiopian region, why not a single emigrant extended its range southwards across the land connection of Abyssinia and Arabia as Africa became united to Euro-Asia; although some of "the most prevalent types of modern African zoology" migrated northwards

[1] Many "arctic" animals may have become so specialized by dwelling in areas subjected to the severe climates, directly south of the ice-fields, during the continuance of glacial conditions, and followed the ameliorating climate northwards, that being the best suited to their conditions of existence, owing to their protracted residence within its influence. The nearly uniform climatic influences which have prevailed over temperate and arctic lands for sixty thousand years are quite sufficient to explain the modification of the fauna and flora subject to those influences, without invoking a purely legendary exodus from the north of boreal forms and their ultimate return. As Dr. Wallace himself writes (*Island Life*, new edition, page 231): "We are sure that some species would become modified in adaptation to the change of climate more readily than others, and these modified species would therefore increase at the expense of others not so readily modified; and hence would arise on the one hand extinction of species, and on the other the production of new forms."

across that newly-formed land and entered Euro-Asia, where, as we know, their remains abound in the Miocene deposits of Greece and elsewhere. This Law of Dispersal also renders it impossible for any "old South Palæarctic fauna" to have "poured into Africa," and to have "finally overran the whole continent" (Wallace, *Geographical Distribution of Animals*, vol. i., p. 288)—a temperate fauna entering a tropical area! We need no "persistence through long epochs of barriers isolating the greater part of Africa from the rest of the world," as Dr. Wallace insists, to account for the absence of such groups as bears, moles, camels, deer, goats, sheep, or such genera as *Bos* and *Sus*; a law forbidding the southern emigration of such types is sufficient to explain the facts, without invoking more or less hypothetical geographical obstructions.

Many of these facts undoubtedly point to a much greater land surface in the Southern Hemisphere as well as to the former existence of a vast Antarctic land—a South Polar continent or continents with extensive land connections stretching far to the north. So long as this Antarctic continent is ignored we shall never arrive at a sound conclusion on the grand question of geographical dispersal. The South Pole must have been the centre of a great and dominant avifauna, of which the Charadriidæ, the Anatidæ, and the Procellariidæ (all groups in which migratory habits are dominant) were strongly-marked features. The great changes which that vast region has undergone and which have utterly exterminated (so far as we know) the avifauna dwelling within it, has been the cause of a great northern dispersal or range contraction southwards, especially of such families as those just mentioned. It is chiefly amongst such species that the avifaunas of Euro-Asia and North America show any similarity. The more temperate and southern forms are utterly dissimilar, and could never have occupied a common North Polar area. I would also allude to that wonderful flora of the Cape area in South Africa, as an example of a range base which has preserved so many species from that vast extermination consequent on the glaciation and submergence which is now characteristic of the Southern Hemisphere. We have only to have a continuous land connection and the withdrawal of glacial conditions as we absolutely did have in the Northern Hemisphere to start that flora on its southern progress, not as an example of species driven north by adverse conditions in South Polar areas returning to old habitats, but as an instance of surviving relics saved from extermination entirely because their base was beyond the limits of the glacial invasion and its attendant submergence; extending their area under a return of favourable conditions for range expansion. Here we have no trace of emigration or range expansion *northwards;* although an emigration of North Temperate forms southwards has been repeatedly invoked by some of our greatest biologists. The whole subject does not come within the scope of the present work, but it is impossible to ignore these facts in dealing with the emigrations and migrations of British birds.

The next chapter will attempt to give an outline of this Glacial Range contraction and Post-Glacial emigration as they are unquestionably indicated by the present distribution of British species and their allied forms.

CHAPTER III.

THE GLACIAL RANGE CONTRACTION AND POST-GLACIAL EMIGRATION OF BRITISH BIRDS.

A New Law of Dispersal—The Third Cold Period of the Glacial Epoch—Probable Avifauna of Refuge Area I. during this Period—Effects of Cold Period on the Charadriidæ—Inter-polar Migration and Emigration—Avian Characteristics of Refuge Area I.—Avifauna of Refuge Area II.—Resident British Species—Southern Representative Forms—Species Resident in British Isles and in Refuge Area II.—British Species that Resort to Refuge Area II. in Winter—Winter Visitors to the British Isles that also Winter in Refuge Area II.—Summer Visitors to the British Isles—Winter Quarters of these in Refuge Area II.—Ancient Sahara Sea a Bar to Emigration from the South—Birds ranging further South in East Africa than in West Africa—Winter Quarters of British Summer Migrants in Refuge Area III.—Circuitous Routes of Migrants to West Africa—British Summer Migrants from the South-East—British or West European Species that have Emigrated from South-Eastern Areas—Their Absence from Iberia—Their Allied Forms and Representative Species—Influence of Competing Species—Reasons for their Absence from British Area—Abnormal Migrants to British Area—West European Species normally Absent from British Area—Reasons for such Absence—Past Emigrations of Storks and Red-crested Pochard—Emigrations of Blue-headed Wagtail and Allies—Situation of British Area now unfavourable to Emigration—Instances showing Impassable Nature of a Sea Barrier—Avian Emigration to Greenland—Nearctic Emigration—Post-Glacial Emigration of Birds in West Europe—Emigration to Iceland and Greenland across the British Area—Table of Emigrants.

BEFORE entering into the details of the subject of this chapter, it may be advisable to promulgate the following

Law, which the reader must repeatedly bear in mind in dealing with the facts presented.

LAW OF DISPERSAL.—Northern Hemisphere species never increase their Breeding Area Southwards—always North, East, or West from a Range Base or Centre of Dispersal, which contained the sole surviving portion of the species during glacial ages. Inter-polar and Inter-hemisphere species north of the Equator never extend their breeding range Southwards if suitable areas are open to the North; Inter-polar and Inter-hemisphere species south of the Equator never extend their breeding range Northwards if suitable areas are open to the South. The tendency of Birds, if not of all living organisms, is to spread in the direction of the Poles. During present time a bird in the Northern Hemisphere never increases its range in a *Southern direction;* it may do so North, North-east or North-west, East or West; but that it never does so South is proved by the fact that no Migration Route is known which trends *South* in spring, or which trends *North* in autumn. During present time a species in the Southern Hemisphere never increases its range in a *Northern direction;* it may do so South, South-east or South-west, East or West, but that it never does so North is proved by the fact that no Migration Route is known which trends *North* in spring, or which trends *South* in autumn. The route of Present Migration is always an unerring guide to, and unfailing indication of, the general line of Past Emigration.

From this Law we draw the following Corollaries:

COROLLARY I.—Migratory birds winter at the present time in the Glacial Refuge Areas or Range Bases occupied by the species or its ancestral forms, or the remnants of such that escaped extermination during the Glacial Epoch, however near or remote such Range Bases or Refuge Areas may be from their present breeding grounds.

COROLLARY II.—The northern limit of a Northern Hemisphere species in any area marks the limit of Emigration from a more southern area, not necessarily south of it. A bird's Breeding Grounds are always at the limits either of its Past Emigrations or Present Migrations.

COROLLARY III.—The Present Migration of a species is a recapitulation of the Past Emigrations or Range Expansion of that species.

COROLLARY IV.—Species never "retreat" from adverse conditions. If overtaken by such they perish, or such portion of the species that may be exposed to them.

COROLLARY V.—Extension of Range is only undertaken to increase Breeding Area, and therefore during most favourable climatic conditions.

COROLLARY VI.—Contraction of Range is only produced by Extermination amongst sedentary species, and probably also by Extermination (through inability to rear offspring) amongst migratory species that are neither Inter-polar nor Inter-hemisphere.

In my opinion this Law will explain not only why no typical Nearctic species encroach upon the Palæarctic Region across Behring Strait,[1] but also assist in solving

[1] I am fully aware that there is a slight intermixture of Palæarctic and Nearctic species in Alaska and North-eastern Siberia, especially amongst migratory birds. I have already alluded to this fact in the *Migration of Birds* (p. 234); whilst Mr. Seebohm has recently pointed out other instances (conf. *Ibis*, 1894, p. 293). There is, however, not the slightest evidence to suggest that any

many puzzling problems of Geographical Distribution elsewhere, especially in the Oriental and Australian Regions. Its bearing on Mammalia, etc., during the Glacial Epoch has already been shown; whilst its relation to the subject of the permanence of continents is equally apparent. Migration primarily is the result of a vast extermination of Birds from their ancient home in the north (or south), caused by the adverse climatic conditions resulting from a glacial epoch, and the constant endeavour of what we must now regard as but the relics of such exiled life to regain and re-people the area that it once occupied during Pre-Glacial time. Whether the climate of Europe was ever so mild and genial during each or any successive warm inter-glacial period, as it unquestionably was in Pliocene ages before the advent of the Glacial Epoch, appears not to be known with absolute certainty. There can, however, be no doubt whatever that towards the close of the Pleistocene Period an intensely cold climate succeeded a warm and genial one. The coming on of this third Cold Period marked the beginning of that last great

one of these species has extended its area either into Alaska or into Siberia in a southerly direction or to a lower latitude than its point of entrance into each region respectively. The facts are in strict accordance with our Law of Dispersal, and suggest a former greater land extension between this portion of the two continents, which we know once existed, and of which the Aleutian Islands are the principal surviving relics. This intermixture of Asiatic and American forms must have taken place before the two continents were separated by the sea. I may also say that we remark no southern extension of these encroaching Palæarctic species into America during winter, or a similar extension of Nearctic species into Asia at the same season. Each migratory species returns to its base of extension in autumn in absolute compliance with the law of its dispersal.

extermination of birds in the Arctic and North Temperate regions; the passing away of that ungenial climate marked the commencement of their return—the beginning of that vast movement northwards of birds which even at the present day has not ceased, as will be seen in a future chapter.[1] These vast changes of climate, there is every reason to believe, took place almost imperceptibly during the course of ages. As the cold climate came on which was eventually to culminate in the third glacial period, the range of the temperate fauna and flora was gradually contracted south by extermination, and that of the "Arctic" fauna and flora slowly followed. Each recurring winter waxed colder and colder, possibly so slowly that no single generation of species detected the change or were perceptibly influenced by it, yet gradually the temperate range of birds became less and

[1] Before the Glacial Epoch finally passed away several severe fluctuations of climate occurred, culminating in one more Glacial Period (the fourth), designated by Professor Geikie as the epoch of the Great Baltic Glacier. During this period Iceland, most of North Scotland, and the highest mountains of England, Wales, and Ireland, together with nearly the whole of Scandinavia and Finland, the Alps, and parts of the Caucasus and the Pyrenees, were glaciated. All living things within glacial influence must have perished during this era, but the extermination was on nothing near so vast a scale as that which characterized the glacical periods preceding it. That this fourth glacial period was a severe check to the emigration of species in certain directions is unquestionable, and its influence in such range extension continues to be demonstrated by the distribution of birds in Western Europe, as we shall eventually learn. The student must bear in mind that the later and less intense periods of the Glacial Epoch successively affected smaller and smaller areas of country—a fact of little or no consequence so far as concerns our general argument, but of importance in studying the present geographical distribution of species.

less northerly; the Arctic area of distribution more and more contracted, until the glaciers and the snow-fields had killed or reduced the range of all living things as low in West Europe as say the 53rd parallel of latitude. England, say from Yorkshire southwards, Northern France, Belgium, Holland, and parts of North-western Germany, eventually became to all intents and purposes the Arctic regions of Western Europe, with a scanty resident avifauna, and a flora in which Arctic willows and mosses, dwarf birches, lichens, and saxifrages were perhaps the most characteristic features.

So far as birds are concerned, and to them, of course, our inquiry is exclusively limited, I am compelled to reject the hypothesis that any temperate species survived the Ice Age as residents in the south of England. I do so because I cannot trace a single species fairly classed as temperate whose winter range to-day in Western Europe reaches no further south than the British Area, or say the northern portion of Range Base or Refuge Area I. It will be interesting to sketch, if possible, the resident avifauna of this district during those far-off ages, when the greater part of England was reduced to an Arctic waste. Then, as now, the few resident birds of the Arctic regions were all Nomadic Migrants—those restless wanderers of the avian world, with no regular seasons of passage, that linger in their accustomed haunts so long as they can obtain food, and that travel no further than to districts where their wants can be supplied. What we now class as Nomadic Migrants were species whose southern range base just escaped the glacial invasion; they are species that survived on the very margin of the ice-sheets and snow-fields, and were the first to move north with an amelioration of climate. In the accompanying table I have arranged those species which probably composed the avifauna of this area during the third, and most probably the second cold periods of the Glacial Epoch, giving their normal southern winter limit in Western Europe at the present time, together with the food on which they habitually subsist, and which is of precisely that character which would be obtainable under the glacial conditions of the climate of that remote period.

SPECIES.	PRESENT S. WINTER LIMIT IN W. EUROPE.	FOOD.
Linota flavirostris ...	Normally France.	Seeds and insects.
,, linaria	France.	,, ,, ,,
Plectrophenax nivalis	,,	,, ,, ,,
Hierofalco caudicans	,,	Lemmings, etc., birds.
,, islandus	British Islands.	,, ,, ,,
,, gyrfalco	,, ,,	,, ,, ,,
Nyctea nyctea ...	Northern France.	,, ,, ,,
Lagopus albus ...	British Islands.	Buds, leaves, seeds, insects, and berries.
Anser brachyrhynchus	English Channel.	Grass, shoots of herbage.
Bernicla leucopsis	,, ,,	Marine grass; crustaceans.
,, brenta	Normally English Channel.	Grass wrack and laver.
Harelda glacialis ...	English Channel.	Mollusks, crustaceans, and water plants.
Somateria mollissima	,, ,,	Mollusks, crustaceans, etc.
,, spectabilis	,, ,,	Mollusks, crustaceans, etc.
,, stelleri ...	,, ,,	Mollusks, crustaceans, etc.
Pagophila eburnea...	English Channel.	Marine animals; seal-droppings, etc.
Mergulus alle	,, ,,	Small crustaceans, etc.
Alca impennis ...	British Islands.	Marine animals, fish.
Uria grylle	English Channel.	Crustaceans, fry, small fish.
Colymbus glacialis	Abnormal below English Channel.	Fish.
Fulmarus glacialis ...	Bay of Biscay.	Mollusks, cuttle-fish, etc.

There seems little doubt that the species mentioned in this table were able to live during these last cold periods of the Glacial Epoch on our southern coasts—then extending much further south towards the Bay of Biscay—or in suitable inland districts as far north as our first Refuge Area extended. Many of these species, it will be seen, were aquatic birds, able apparently to subsist anywhere near to open water; whilst the others were hardy species living on Arctic berries, seeds, shoots, and buds, and on their smaller and more helpless companions, as well as on the various animals that we know also survived the glacial invasion of the land. Trees were entirely absent from the British or most northern portion of this Refuge Area, or were too small and stunted for the requirements of Woodpeckers (PICIDÆ) and other arboreal species. Hence the non-migratory habits of these birds. It may be remarked that every one of the birds included in this table could have lived in winter in such an area; if they could not they would assuredly have vanished for ever, as we have uncontrovertible evidence to show that the food on which they are known now to subsist was then actually obtainable.

As further confirmation that these birds formed the avifauna of this area during the Ice Age, we may mention that not a single species is represented in the south by an allied race. Broadly speaking, then, not one of these species dwelt south of our first Refuge Area, for not one down to the present day normally wanders south of that area, and not one is represented south of it by a closely allied form which would indicate a southern Range Base and contraction of range by extermination during the Glacial Epoch. Very different, however, was the case of the Arctic birds that lived on animal substances alone—the Plovers, Sandpipers, and their allies. The range of these species at the coming on of the Ice Age became more and more southerly, and their migrations inter-polar, probably because they could not find suitable winter quarters or breeding grounds except in the Polar regions. As I have already shown in the *Migration of Birds*, we have abundant proof of this Inter-polar Migration and Emigration, not only in the vast journeys many of these birds still undertake, but in the many allied forms left behind in the Southern Hemisphere when the northern Glacial Epoch passed away and the North Polar breeding grounds became available once more. It is a most significant fact that every Charadriinæ species goes further south to winter than our first Refuge Area; that is to say, that although some members of certain species may winter therein, other members extend their flight to the south of it. Not a single species of the Charadriidæ can be classed as a Nomadic Migrant. *Not a single migratory species* (non-Nomadic) *throughout the Northern Hemisphere winters exclusively within what are defined by uncontrovertible evidence to be the limits of glaciation*—eloquent testimony, I take it, of the southern range bases during Pleistocene time of all the surviving species that have now re-peopled the once glaciated land areas. To my mind these facts are convincing proofs that the high Polar latitudes were deserted by bird-life ages before it was exterminated from more temperate and southern latitudes; in other words, that the various climatic fluctuations of the Ice Age waxed and waned very slowly. It should be remarked that the range of some of the species tabulated may be lower in the New World than in Europe, due to more rigorous climatal conditions and to the much greater southern extension of the glaciers in North America.

It requires but little strain upon the imagination to recall the avian characteristics of this Refuge Area I. at the climax of the third cold period of the Ice Age. England south of Yorkshire, the Bristol Channel, and much of the land now lying submerged beneath the English Channel and the North Sea, were probably in the condition of an Arctic tundra, bounded by the sea on the south in which huge icebergs floated, and on the north by the vast glacier whose southernmost slopes retreated or advanced a little way as summer or winter came on. The resident land birds were few in number. The Twite and the Mealy Redpole wandered about the winter wastes subsisting on the seeds that were obtainable amongst the snow ; the charming little Snow Buntings gathered into flocks and led the nomad life they still continue to lead, going no further south than their very limited range base extended, hurrying north again with the first dawn of spring. The Willow Grouse, clad in its winter plumage of unsullied white, managed to find enough food in the buds and seeds and twigs, and in the frozen Arctic ground fruits which it managed to obtain by burrowing into the snow. The large Arctic Falcons and the Snowy Owl kept closely to the tundras and the shores, preying upon the Finches and the Grouse and the various species of water birds that no glacial invasion appears to have succeeded in utterly exterminating from this area. The Pink-footed Goose, the Bernacle Goose, and the Brent Goose lingered at their range base throughout the winter in this Refuge Area ; the Harlequin Duck and the Long-tailed Duck, together with the three species of Eider, if their present distribution be any indication, apparently did the same. The Ivory Gull survived amongst the icebergs and the floes ; as also did the Great Auk, the Little Auk, and the Black Guillemot ; whilst the White-billed Diver and the Fulmar seemed to have ranged no further south than the open water, which probably then existed all the winter through along the southern coasts of this Refuge Area, owing to the ameliorating influence of the Gulf Stream.

Of the summer aspects of this region it is difficult to speak with any degree of certainty. There can, however, be little doubt that as the climate moderated numbers of species would come north into this area to breed, although they had imperatively to retire south to winter in Refuge Area II. The Charadriinæ birds that had been banished to the South Polar area would also gradually return as breeding species, but they were possibly among the latest to do so, waiting until at least some portion of the higher latitudes of Europe was free from ice, or the glaciers at the South Pole began to contract their area. The phenomenon of Migration in Western Europe as we see it to-day was undoubtedly initiated with the passing away of the third glacial period. Species after species became more and more southerly as the winters almost imperceptibly waxed longer and colder, and exterminated all the northern portion of the sedentary species; slowly the northern breeding range of species after species became more and more contracted as the food supply decreased, or the summer temperature lowered. And so matters went on until only our three Refuge Areas contained representatives of the species that had been exterminated, or whose northern range had been contracted. For the most part these three areas were probably inhabited by a sedentary avifauna; but even at the climax of this cold period I think there can be little doubt that a certain percentage of the temperate species in Refuge Area II.—the hardiest—undertook a migration from south to north with each recurring summer. With a great number of these hardy species Migration finally ended in Emigration, as the climate ultimately became sufficiently genial for them to winter in safety. There may also have been a considerable local migration within each respective area, north in spring, south in autumn, gradually extended as the glacial conditions passed away; for to my mind it is difficult to believe that birds did not respond in a migratory way from the south to the changes of the seasons with their accompanying advantages or perils.

Post-Glacial Migration therefore must have originated south of our area, and gradually extended north as the range of species expanded in obedience to more genial conditions, and in many cases have lapsed as the climate moderated.

Now with regard to the birds that reached Refuge

RANGE BASE OR REFUGE AREA II

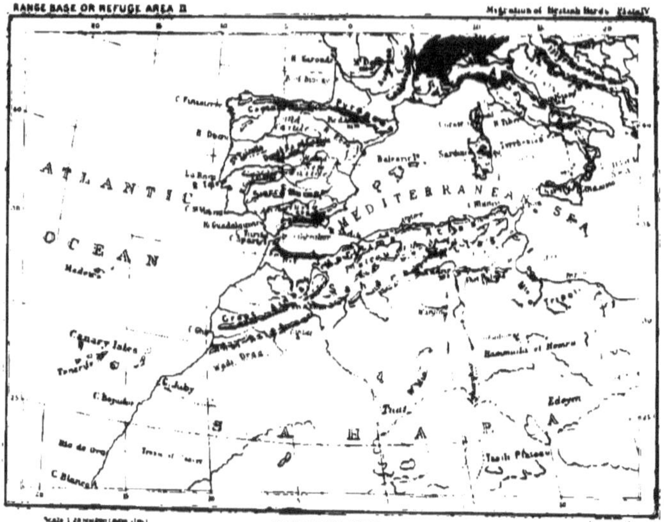

CHAPMAN & HALL, LIMITED

Area II. There can be no doubt whatever that during the glaciation of the British Area, the range of a very large number of the then indigenous species was contracted down to the Iberian Peninsula, and what is now North-west Africa and the Canaries, and that from this area the survivors gradually emigrated northwards as the ice retreated. We find absolute proof of this fact in the traces of that ancient range contraction and subsequent emigration which are still preserved to us in the by no means small number of representative species and races of British birds left behind, especially in the southern portions of this second Refuge Area, when the movement north commenced. At the present time our resident species may be computed at about 115.[1] It is a profoundly interesting and significant fact that of these no less than twenty-one species, or more than one-sixth of the whole, are represented in Iberia, North-west Africa, and the Canaries by closely allied species (in some cases two or three distinct representatives) or sub-specific forms whose complete segregation may be retarded or prevented by their interbreeding with the individuals of the typical species which reach that area as winter migrants, not necessarily from the British Islands, but from continental Europe. These species, with their representative forms or species, are shown in the accompanying table.

[1] Probably this is much in excess of the actual number, as many species are largely increased in number by migratory arrivals from northern or eastern areas, and our own breeding individuals may draw south, seeing that many species supposed to be resident with us are well-known winter migrants further south.

BRITISH RESIDENTS.	REPRESENTATIVE SPECIES OR RACES.		
	IBERIA.	N. W. AFRICA.	CANARY ISLANDS.
Falco peregrinus		Falco barbarus	
Buteo vulgaris		Buteo desertorum	Buteo desertorum (?)
Asio brachyotus	Asio capensis	Asio capensis	
Corvus corax		Corvus tingitanus	Corvus tingitanus
Pica caudata	Pica caudata mauritanica	Pica mauritanica	
Garrulus glandarius		Garrulus cervicalis	
Fringilla chloris	Fringilla chloris aurantiiventris	Fringilla chloroticus	
Fringilla cœlebs		Fringilla spodiogena	Fringilla tintillon
Lanius excubitor[1]	Lanius meridionalis	Lanius algeriensis	Lanius algeriensis
Parus ater		Parus ledouci	
Parus cæruleus		Parus ultramarinus	Parus ultramarinus, P. teneriffæ, P. palmensis, P. ombriosus
Acredula rosea	Acredula irbii		
Regulus cristatus			Regulus teneriffæ
Cinclus aquaticus	Cinclus aquaticus albicollis	Cinclus minor	
Erithacus rubecula[2]			Erithacus superbus (Teneriffe and Grand Canary)
Pratincola rubicola			Pratincola dacotiæ
Anthus pratensis			Anthus bertheloti
Gecinus viridis	Gecinus viridis vaillantii *et* Gecinus sharpii	Gecinus vaillantii	
Picus major		Picus numidicus	Some differences
Picus minor		Picus ledouci	
Larus argentatus	Larus argentatus cachinnans	Larus argentatus cachinnans	Larus argentatus cachinnans

[1] This species cannot fairly be classed as a resident now in the British Islands, but is one which does not pass to the south of the Pyrenees to winter.

[2] Mr. Scott Elliot has recently recorded this species from an elevation of 10,000 feet, near the Albert Edward Nyanza, Equatorial Africa (conf. *Nature*, 1895, p. 271).

This list, however, does not quite exhaust the number of representative species or forms that have been thus isolated, and more or less perfectly segregated from

British resident species. For in the Italian peninsula and on some of the various islands of the West Mediterranean—which during the Glacial Epoch were not islands at all, as a glance at the map of this Refuge Area will show—there are other local species or races. None of these islands are yet thoroughly explored, and probably other forms still remain to be discovered. As may naturally be expected, the representative species or forms in this area are neither numerous nor striking, due partly to the enormous amount of migration taking place over that area, which tends to check segregation by the intermixing of individuals, and to its far less complete isolation from continental Europe. The most characteristic instance known to me is the island form of the Nuthatch (*Sitta cæsia*), confined so far as has yet been ascertained to Corsica, and known as *Sitta whiteheadi.* A second instance is *Passer italiæ*, the Corsican and Italian representative of the Common Sparrow (*Passer domesticus*).

In a considerable number of the instances tabulated above there can be little if any doubt that the southern races or species are the oldest, the parent races, the northern forms being of more recent segregation, because the young in first plumage of the latter resemble the adults of the former—recapitulate in this portion of the course of their development the stages through which the species has passed in its evolution. I have been unable to get the requisite particulars in some cases, but in the following instances confirmation of the fact is forthcoming.

PARENT SPECIES. ADULT.	NORTHERN DEVELOPMENT. YOUNG.
Buteo desertorum. Smaller: tail bars nearly obsolete.	*Buteo vulgaris.* Smaller: tail bars less in number.
Corvus tingitanus. Smaller: breast hackles wanting.	*Corvus corax.* Smaller: breast hackles wanting.
Pica mauritanica. Smaller: no light rump patch.	*Pica caudata.* Smaller: rump patch much less in extent.
Fringilla spodiogena. Underparts much paler.	*Fringilla cœlebs.* Underparts much paler.
Parus ledouci. Nape and cheeks yellow.	*Parus ater.* Nape and cheeks yellow.
Acredula irbii. No rosy tinge on scapulars, which are grey.	*Acredula caudata* et *rosea.* No rosy tinge on scapulars.
Anthus bertheloti. Smaller: underparts narrowly striated: no green tints in plumage.	*Anthus pratensis.* Smaller: very buff in general coloration: striations on underparts smaller.
Gecinus viridis vaillantii et *sharpii.* No black on forehead and round eye.	*Gecinus viridis.* Black absent from ear coverts, lores, and round eye.
Picus numidicus. Scarlet on breast.	*Picus major.* Occasional reversions to red on breast.
Picus ledouci. Underparts strongly suffused with brown.	*Picus minor.* Underparts not so pure in colour.

Equally suggestive is the fact that of the remaining species no fewer than 64, or nearly two-thirds of the entire number of resident British species, are still resident in the Iberian Peninsula, in the Balearic Islands, in Corsica and in Sardinia, North-West Africa, or the Canaries, and do not present any differences from the northern individuals of the same species. Their distribution is continuous throughout the area, and has probably always been so during the entire period over which our present investigation extends, their range contracting southwards by extermination and expanding northwards by colonization contemporaneous with, and in obedience to, the changes of climate. The following table will serve readily to demonstrate these interesting facts.

RESIDENTS IN BRITISH ISLANDS.	IBERIA.	N. W. AFRICA.	CANARY ISL.
Aquila chrysaëtus	×	×	
Milvus regalis	×	×	
Accipiter nisus	×	×	
Asio otus	×	×	
Syrnium aluco	×	×	
Corvus corone	×		
[1] Corvus cornix			
Corvus monedula	×		
Pyrrhocorax graculus	×		×
Loxia curvirostra	×		
Pyrrhula vulgaris	×		
Linota cannabina	×	×	×
[2] Fringilla carduelis	×	×	×
Coccothraustes vulgaris	×	×	
Passer domesticus	×	×	
Passer montanus	×	×	
Emberiza miliaria	×	×	×
Emberiza cirlus	×	×	
Emberiza citrinella	×	×	
Alauda arvensis	×	×	
Alauda arborea	×	×	
Certhia familiaris	×		
Sitta cæsia	×		
Parus cristatus	×		
Parus major	×	×	
Panurus biarmicus	×		
Sylvia provincialis	×	×	
Turdus viscivorus	×	×	
[3] Turdus musicus	×	× ?	
Merula merula	×	×	
Accentor modularis	×		
Troglodytes parvulus	×		

[1] Resident in Balearic Islands, Corsica, Sardinia, Italy, and Sicily.

[2] Recorded by Mr. Scott Elliot from an elevation of 10,000 feet (Mt. "Ruwenzori") near the Albert Edward Nyanza (conf. *Nature*, 1895, p. 271).

[3] This species now only reaches the Canary Islands, the extreme southern portion of its refuge area in Western Europe, as a winter visitor in varying numbers, having ceased to breed therein.

RESIDENTS IN BRITISH ISLANDS.	IBERIA.	N. W. AFRICA	CANARY ISL.
Alcedo ispida ...	×		
[1] Anser cinereus ...	×		
Tadorna cornuta ...	×		
Anas boschas ...		×	
Anas clypeata	×	×	
Anas crecca ...	×		
Anas strepera ...	×		
[1] Fuligula ferina ...	×		
[2] Vanellus cristatus	×	×	
[2] Totanus calidris	×	×	
[1] Tringa alpina ...	×		
Scolopax rusticula		×	
Otis tarda	×	×	
Botaurus stellaris	×	×	
Phalacrocorax carbo ...	×	×	
Phalacrocorax graculus	×	×	
Uria troile	×		
Puffinus anglorum ...	×		
Procellaria pelagica ...	×		
[2] Larus fuscus ...	×		
[1] Larus ridibundus	×		
Podiceps cristatus	×		
Podiceps minor ...	×		
[2] Gallinula chloropus ...	×		
[2] Fulica atra... ...	×		
Rallus aquaticus	×		
[2] Crex bailloni	×		
Columbus palumbus ...	×	×	
Columbus œnas	×		
Perdix cinerea ...	×		
Tetrao urogallus	×		
Lagopus mutus	×		

[1] These species only occasionally breed as far south as Iberia; the habit having nearly lapsed owing to the Post-Glacial northern extension of range.

[2] These species now only reach the Canary Islands, the extreme southern portion of their refuge area in Western Europe, as winter visitors in varying numbers, having ceased to breed therein.

Continuing our investigations still further we find that out of the resident birds, or species, to be met with in some part of the British Islands throughout the year, no less than 29 resort to Refuge Area II. in winter, indicating a former range base, many of them there can be little doubt being individuals from our islands whose place in them at that season is taken by other individuals visiting us from more northerly or easterly areas, and which regard our islands as their refuge because it is the area from which they emigrated north or east to colonize other lands in past ages. These species are indicated in the following table.

SPECIES.	IBERIA.	N. W. AFRICA.	CANARY ISL.
Falco tinnunculus ...	×	×	×
Falco æsalon	×	×	
Circus cyanus	×	×	
Haliaëtus albicilla ...	×	×	
Sturnus vulgaris	×	×	
Motacilla yarrellii ...	×	×	
Fringilla spinus	×	× rare	
Corvus frugilegus	×	×	
Ardea cinerea	×	×	
Sula bassana	×	×	
Procellaria leachi	×	×	
Anas acuta	×	×	
Anas penelope	×	×	× rare
Fuligula cristata	×	×	
Fuligula nigra	×	×	
Mergus serrator	×	×	
Mergus merganser	×	×	
Podiceps cornutus	×	×	
Colymbus septentrionalis ...	×		
Alca torda	×	×	×
Numenius arquata ...	×	×	×
Limosa melanura ...	×	×	×
Totanus glottis	×	×	×
Scolopax gallinago ...	×	×	×
Hæmatopus ostralegus	×	×	×
Larus canus...	×	×	×
Larus tridactylus	×	×	×
Stercorarius richardsoni ...	×	×	
Stercorarius catarrhactes ...	×	×	

Passing from the species that are resident in our islands, or found in them throughout the year, we shall still find the same significant facts presented by those species that are only observed in our area as Winter Visitors. Of these regular winter migrants individuals of the following tabulated species prolong their flight southwards to Iberia, North Africa, and the Canaries, which indicates an ancient range base of those species and where such species were preserved from extinction in Western Europe during the glaciation of northern lands, and from which they started as emigrants on their northward extension with the change of climate.

SPECIES.	IBERIA.	N. AFRICA.	CANARY ISL.
Archibuteo lagopus	×		
Fringilla montifringilla	×	×	
Turdus pilaris	×	×	×
Turdus iliacus	×	×	
[1] Charadrius helveticus ...	×	×	×
Strepsilas interpres ...	×	×	
Calidris arenaria	×	×	
[1] Limosa rufa	×	×	
Scolopax gallinula	×	×	
Phalaropus fulicarius	×	×	
Tringa maritima	×	×	
Tringa canutus		×	
Stercorarius buffoni			
Stercorarius pomatorhinus ...		×	
Larus glaucus		×	
Larus minutus	×		
Cygnus musicus			
Cygnus bewicki			
[2] Cygnus olor			
Anser segetum	×		
Anser albifrons	×		
Mergus albellus	×		
Fuligula fusca	×		
Fuligula marila	×		
Clangula glaucion	×	×	
Botaurus stellaris	×	×	×

[1] It is a very remarkable fact that both these species do not go south of the Equator in Africa, but individuals of the former prolong their migrations in Asia as far south as Australia, and of the latter as far south even as New Zealand! The Black-tailed Godwit and Common Sandpiper are other instances.
[2] It is only in winter that this species visits us in a thoroughly wild state.

There are five other species which might almost be included in the above table of our winter migrants; but there can be little doubt that the Goshawk (*Astur palumbarius*), the Black Redstart (*Ruticilla tithys*), the Firecrest (*Regulus ignicapillus*), the Little Bustard (*Otis tetrax*), and the Red-necked Grebe (*Podiceps rubricollis*) are only abnormal winter visitors to the British Islands, or, what is much more likely, may have once bred there and become exterminated (*conf.* pp. 183—186). If this be not a true interpretation of the facts, we are confronted with an insurmountable difficulty of Distribution, for we should find the anomaly of five species breeding and wintering south of us in Iberia, North-west Africa, and elsewhere, but wintering only in our area. The mild southern counties of our country enable individuals of these species to winter at or near the northern limit of their distribution. There can be little doubt that the two Passerine species would ultimately become residents again in our area if left unmolested; the three larger species would possibly become so if our islands were not so thickly populated, or the individuals that reach us were not so promptly exterminated. I may here also take the opportunity of remarking that we have not a single species wintering in the British Islands and South-eastern Europe, and absent from South-western Europe and North-western Africa (although many of course winter in both), which is a most suggestive fact, indicating the sources of our avifauna and its Refuge Areas or Range Bases during glacial times.

When we come to deal with the Summer Visitors to the British Islands, the vast importance of Refuge Area II. is demonstrated in no uncertain way. This Refuge Area continues down to the present time to be the winter home of almost every bird that migrates to our country in spring to breed. The following table indicates the species and the precise winter habitat of each, or, in cases of wider dispersal, of the individuals of such species that breed in Western Europe.

	SPECIES.	WINTER AREA.
	Falco subbuteo	Iberia : breeds and winters.
	Pernis apivorus	Iberia (breeds and winters) ; North and West Africa.
A	Circus cineraceus	Iberia (breeds and winters) ; N. Africa.
	Pandion haliaëtus	Iberia ; N. Africa : breeds and winters.
A	Muscicapa atricapilla	Iberia ; N. Africa (few breed in Algeria).
	Motacilla alba ...	Iberia (few breed) ; N.W. Africa.
	Anthus arboreus	Iberia : N.W. Africa.
	Merula torquatus	Iberia (breeds and winters) ; N.W. Africa.
B	Erithacus luscinia	N.W. Africa : few breed.
B	Ruticilla phœnicurus ...	N.W. Africa.
	Pratincola rubetra	N.W. Africa.
B	Saxicola œnanthe	N.W. Africa : few breed.
	Acrocephalus phragmitis	N.W. Africa.
B	Acrocephalus arundinaceus	N.W. Africa.
A	Sylvia hortensis	N.W. Africa.
	Sylvia curruca	Iberia ; N.W. Africa.
A	Sylvia cinerea	N.W. Africa.
	Sylvia atricapilla	Iberia ; N.W. Africa.
	Locustella locustella ...	Iberia ; N.W. Africa.
	Locustella luscinoides ...	Iberia ; N.W. Africa.
A	Phylloscopus sibilatrix ...	N.W. Africa.
B	Phylloscopus trochilus ...	Iberia ; N.W. Africa.
B	Phylloscopus rufus	Iberia ; N.W. Africa.
B	Upupa epops	N.W. Africa ; Canary Islands.
A	Iynx torquilla	N.W. Africa.
A	Crex pratensis	N.W. Africa.
	Crex porzana	Iberia ; N.W. Africa.
B	Coturnix communis ...	Iberia ; N.W. Africa.
	Botaurus minutus ...	N.W. Africa.
	Platalea leucorodia ... ... (*extinct as a breeding species*).	N.W. Africa.
B	Grus communis ... ... (*extinct as a breeding species*).	Iberia ; N.W. Africa.
	Œdicnemus crepitans	Iberia ; N.W. Africa.
	Eudromias morinellus	N.W. Africa.
	Ægialophilus cantianus ...	Iberia ; N.W. Africa.
B	Recurvirostra avocetta ... ... (*extinct as a breeding species*).	N.W. Africa.
B	Numenius phæopus...	N.W. Africa ; Canary Islands.
	Limosa melanurus ... ... (*extinct as a breeding species*).	Iberia ; N.W. Africa.
B	Totanus hypoleucus	Iberia ; N.W. Africa.
B	Totanus glareola ... ... (*extinct as a breeding species*).	N.W. Africa.
B	Totanus ochropus	Iberia ; N.W. Africa.
B	Totanus pugnax	N.W. Africa.

SPECIES.	WINTER AREA.
Sterna nigra ... ... *(extinct as a breeding species).*	N.W. Africa.
Sterna cantiaca...	West coasts of Africa.
Sterna hirundo...	West coasts of Africa.
Sterna dougalli...	West coasts of Africa.
Sterna arctica	West coasts of Africa.
Sterna minuta	West coasts of Africa.
Stercorarius richardsoni ...	West coasts of Africa.
Fratercula arctica	N.W. Africa.
B Anas circia	N. Africa.

The species marked A in the above table winter, or winter and breed, in Iberia or North-west Africa, or both, but only winter in or pass through South-east Europe, Asia Minor, Palestine, and North-east Africa on passage to regions further south to spend the cold season, extending in some cases even to the Cape Colony. This circumstance appears to me to prove two facts. First, that the ancient Sahara Sea barred further progress to the south in the western portions of Africa, as it had previously barred northern progress say in Pliocene ages; and second, that North-west Africa and Iberia formed the glacial Refuge Area or Range Base of the British or West European portion of those species; the individuals breeding further north in the eastern portions of the continent than in the western portions, perhaps, during Pre-Glacial ages, and even continuing to do so during the progress of the Glacial Epoch, as we have already seen the extension southwards of the ice-sheets in the east was much less than in the west (*conf.* p. 28), thus leaving a wider expanse of breeding area northwards during the successive cold periods. By applying the well-established Law of Migration, that the birds that breed the furthest north winter the furthest south, we can readily understand why eastern individuals of the same species visit the Cape in winter, whilst western individuals go no further south than North-west Africa and the Canary Islands—their base of northern extension.

The species marked B in the above table are birds that go much further to the south in East Africa than in West Africa; not necessarily because they breed any further north in East Europe or West Asia than in West Europe, but because the passage up or down the Nile valley has always been a more continuous one, whilst in West Africa the Sahara Sea or its now sandy wastes were an insurmountable barrier to emigration. The Terns, being thoroughly oceanic species, do not come within the scope of these remarks. The evidence also seems to suggest that some of these migrants, after going down the Nile valley, spread westwards across the Soudan, even to the Atlantic sea-board, as if they were following the southern coast-lines of the ancient sea!

There is another small group of Summer Migrants to the British Islands which appears to me to have unquestionably dwelt in Refuge Area III. during the Glacial Epoch. The winter quarters of these birds may be either in the Mediterranean basin or in Tropical Africa. They appear to reach those districts by way of Iberia, thence eastwards through Algeria and Tripoli, and southwards down the Nile valley. This class of summer migrants illustrates very vividly the intricate paths followed by birds on passage, and demonstrates how in the remote past the colonists gradually extended their range, first eastwards and westwards across the Soudan, round the southern borders of the Sahara Sea or Desert, and then northwards up the valley of the Nile, ultimately following the more recently-formed coast-area of Tripoli, as the sea receded, colonizing Algeria, and ultimately spreading northwards into Iberia, into France, and eventually the British Area. The line of Emigration followed by those species in the past continues to be the route of Migration in the present. A few individuals of some of the species tabulated below may possibly winter in Algeria and Tripoli, as, for instance, the Swallow, but the great majority pass on to Tropical Africa, even to West Africa and the Atlantic sea-board, by this circuitous route! It is a very significant fact that the majority of these migrants are late to arrive at their breeding grounds, as is customary with most if not all species from the south-east. The only exceptions are those of species the British individuals of which may probably winter in Algeria—the three Swallows, and the Yellow Wagtail. It is a still more significant and suggestive fact—confirming the views above expressed—that with the solitary exception of the Turtle Dove not one of the species is common on migration at the Canary Islands, as we should reasonably expect to be the case did the birds journey north from Tropical Africa by this route. Nay more, such common species as the Yellow Wagtail, the Swift, the Goatsucker, and the Gargancy have never been noticed at the islands at all; whilst such wide-spread birds as the Cuckoo, the Sand and House Martins, the Golden Oriole, and the Spotted Flycatcher are only noted as stragglers to the group. The Turtle Dove, however, is a very common summer visitor to the Canaries, but there is no doubt whatever that it reaches the islands from the east, from the Soudan, and not from the south, inasmuch that the bird is unknown in South Africa, and is closely allied to the Central African *Turtur isabellinus*, a regular summer visitor down the Nile valley to North-east Africa. The various species with their winter quarters —once their glacial Refuge Area—are indicated below.

SPECIES.	WINTER AREA.
Muscicapa grisola ...	N. Africa (?) : Tropical Africa, and south to Cape.
Oriolus galbula ...	Tropical Africa, and south to Natal.
Motacilla raii ...	Algeria, Tripoli (?): Tropical Africa, and south to Transvaal.
Hirundo rustica ...	Algeria, Tripoli (?): Tropical Africa, and south to Cape.
Chelidon urbica ...	Algeria, Tripoli (?): Tropical Africa.
Cotyle riparia ...	Algeria, Tripoli (?): Tropical Africa, and south to Transvaal.
Cuculus canorus	Tropical and South Africa.
Cypselus apus	Tropical and South Africa.
Caprimulgus europæus [1]	Algeria, Tripoli (?): Tropical Africa, and south to Natal.
Turtur communis	Tropical Africa.

[1] This species is a very late migrant—a fact which seems to prove that the bird is *not* a winter resident in any part of North Africa.

There is yet another and smaller group of Summer Migrants to the British Islands, descendants of birds that also dwelt in Refuge Area III., and whose route of Migration at the present time indicates the line of Emigration taken by the species in past ages, as the breeding or summer range was increased when the ungenial climatic conditions passed away. They are as follows:—

SPECIES.	WINTER AREA.
Lanius collurio	Greece on passage; South Africa. *viâ* the Nile Valley.
Acrocephalus palustris ...	Nile Valley; Tropical Africa, and south to Natal.
Phalaropus hyperboreus	Black Sea Basin: Persia: India.

These species appear to reach the British Islands by way of the Danube and Rhine valleys. They are also, significantly enough, unknown in the Iberian Peninsula, although two of them, the Red-backed Shrike and the Marsh Warbler, are common enough in France during summer. Mr. Dresser professes to have identified Marsh Warblers from Malaga, but Mr. Howard Saunders, in whose collection they were, very rightly considers them to be nothing but Reed Warblers, *Acrocephalus arundinaceus*. This shows how important a correct knowledge of Migration may be, even in the identification of specimens. These birds are all very late migrants (amongst the last species to appear in spring), as we have already seen is the case with birds from the south-east. All summer birds of passage from the south-east to Western Europe are late migrants; and this fact seems to suggest that the habit was acquired during the passing away of the vast glaciers from Central Europe, when the summers were probably shorter than they are now—the migration then partaking of the character which is the present feature of the phenomenon in the Arctic regions, where the season is very late and very short. Species from the direct south—from North-west Africa and Iberia—had in those remote ages (and as they still continue to have) a much earlier route open to them, enabling them to push northwards sooner in the spring (a habit which is still continued), owing to the ameliorating influence of the Gulf Stream and the milder glacial conditions. It may also be remarked that none of these late migrants from the south-east are dominant or widely dispersed in the British area—the Red-backed Shrike being principally confined to the southern and central counties of England, the Marsh Warbler to one or two of the southern counties only, the Red-necked Phalarope chiefly to the Shetlands, Orkneys, and Outer Hebrides. We shall also find that even with resident birds that have emigrated from the south-east, extremely few are widely distributed or dominant species in our area (*conf.* pp. 88--91).

It will thus be seen that of the 63 species that may be fairly classed as regular summer visitors to the British Islands (or some part of them), no less than 60 species may be said, almost with absolute certainty, to have reached them by way of the Iberian Peninsula and France, and but three species can be reasonably presumed, judging from their present geographical limits, to have reached them by no other way than from the south-east, most probably *viâ* the valleys of the Danube and the Rhine. Could anything testify more eloquently or more significantly to the source whence the British avifauna has been derived?

Whilst dealing with species obviously of south-eastern origin, it may be advisable here to treat with another group of birds resident in or wanderers on nomadic or abnormal flight to the British Islands—birds whose range base during the Glacial Epoch extended to Refuge Area III., or which have gradually extended their range to Western Europe from Asia after the Ice Age had passed away.

SPECIES.	EUROPEAN SUMMER AREA.	WINTER AREA.
Loxia bifasciata	Pine forests of N. Russia.	S. Sweden to N. France and N. Italy; Austria. Poland, S. Russia, and S. Siberia.
Pinicola enucleator ...	Conifer region near Arctic Circle.	S. Norway, Denmark, N. Germany; S. Siberia.
Carpodacus erythrinus	Breeds as far west as the Baltic Provinces.	India and Burma.
Calcarius lapponicus ...	Breeds as far west as Norway.	Mongolia and China.
Emberiza melanocephala	Breeds as far west as Italy.	Western and Central India.
Emberiza pusilla	Breeds as far west as Archangel.	India, Burma, China.
Emberiza rustica	Breeds as far west as the Baltic.	Russian Turkestan, China, Japan.
Otocoris alpestris... ...	Scandinavia, N. Russia.	S. Europe (except Iberia), S. Siberia, S.W. Turkestan, N. China.
Anthus cervinus	Breeds as far west as Scandinavia.	Egypt, Nubia, and Abyssinia.
Lanius minor ...	Breeds as far west as E. France.	Nile Valley to S. Africa.
Ampelis garrulus... ...	Arctic forests.	Central Europe W. to France and E. to Turkey; S. Siberia.
Muscicapa parva	Breeds as far west as Germany.	N.E. Africa, Persia, India, and China.
Erithacus suecica ...	Breeds as far west as Scandinavia.	N.E. Africa.
Sylvia nisoria ...	Breeds as far west as S. Sweden.	N.E. Africa, passage; Central Africa, winter.
Hypolais icterina ...	Breeds as far west as N. France.	S.E. Europe, N.E. Africa on passage; S. Africa, winter.
Falco vespertinus ...	Breeds as far west as Hungary.	S.E. Europe, N.E. Africa, passage; S. Africa, winter.
Surnia funerea	Pine forests of Scandinavia and N. Russia.	N. Germany, Central and S. Russia, S. Siberia.
Nyctala tengmalmi ...	Arctic pine forests, Alps, Carpathians.	N. France, Germany, Cent. Russia, S. Siberia.
Tetrao tetrix	Pine and birch forests of N. and Cent. Europe.	Sedentary.
* Scolopax major... ...	Breeds as far west as Holland.	Few in basin of Mediterranean; S. Africa.
* Tringa temmincki ...	Breeds as far west as Scandinavia.	Few in Iberia, Algeria; majority in Nile Valley; India, Burma, China, Malaysia, etc.
* Tringa minuta	Breeds as far west as N. Norway.	Few in N.W. Africa; bulk in S. Africa, Persia, India, Burma.

SPECIES.	EUROPEAN SUMMER AREA.	WINTER AREA.
* Tringa subarquata ...	Breeding area unknown.	Ethiopian, Oriental, and Australian Regions; few in basin of Mediterranean.
Tringa platyrhyncha ...	Breeds as far west as Scandinavia.	E. Mediterranean basin; N.E. Africa to Madagascar, Meckran coast, N. India, Malay Archipelago, etc.
* Totanus fuscus	Breeds as far west as Scandinavia.	Africa N. of Equator; few S. to Cape, India, Burma, China.

With the few exceptions shortly to be noted, none of these birds are known to visit Iberia normally, no more than they are known to visit us. So far as the British Area is concerned the Black Grouse (*Tetrao tetrix*) is the only normal species, and in this case it is possible that a portion of this species refuged in Area II. during the Glacial Epoch, or so near to it (in the N.W. portions of Refuge Area III.) that they began to emigrate across France and Britain at the earliest favourable moment, yet comparatively late, as the absence of the species from Ireland suggests. The birds, then, tabulated above are all species of Eastern origin, those with closely allied forms or representative species all inhabiting the east and south-east, or, as in the case of the Bluethroat (*Erithacus succica*), for example, they are actually the eastern representative of the West European species, which has succeeded to a great extent in encroaching upon the area of its ally. Thus *Loxia bifasciata* is most nearly allied to the *Loxia leucoptera* which is found across North America from Alaska to Labrador, and perhaps South Greenland; *Pinicola enucleator* with its allied species *P. subhimachalus* inhabiting the Himalayas; *Carpodacus erythrinus* with its several allies in Palestine, the Caucasus, Turkestan, South Siberia, and the Himalayas; *Calcarius lapponicus*, with its nearest allies in North America; *Otocoris alpestris*[1] with its various allies in South-east Europe and South-east and Central Asia, notably *O. bilopha* which inhabits the deserts of Arabia and North Africa; *Ampelis garrulus* with its close ally *A. phœnicoptera* of Japan, and *A. cedrorum* of North America; *Muscicapa parva* with its representative species *M. hyperythra* in India; *Falco vespertinus* with its eastern representative *F. amurensis* confined to East Siberia, East Mongolia, and N. China during the breeding season; *Surnia funerea* with its allies *S. doliata* in Siberia, and *S. nisoria* in North America; *Tetrao tetrix* with its only allied form *T. mlokosiewiczi* inhabiting the Caucasus; and *Tringa minuta* with its closely allied races *T. ruficollis* of Eastern Siberia and *T. subminuta* of Eastern Siberia and Behring Island, leading on to the *T. minutilla* of Arctic America.

I might here take the opportunity of remarking upon the very interesting instance of geographical distribution, as showing the influence of competing species, presented by the Icterine Warbler *Hypolais icterina*. This bird breeds in North France, Belgium, Holland, Denmark, and North Germany, and visits Scandinavia beyond the arctic circle for nesting purposes (conf. *Ibis*, 1894, p. 229), yet, as it is decidedly a south-eastern species, it has not reached our area. It seems, however, to have

[1] In my opinion the utter absence of this species from Iberia is a very conclusive proof that the Post-Glacial emigrations of the Shore Lark started from the South-east. There is no trace whatever, geographically, between *Otocoris alpestris* and *O. bilopha* in the West, but ample evidence (through allied races especially) of their former continuity of area (and community of origin) in the East.

prevented the Iberian and North-West African *Hypolais polyglotta* from extending its range north of the Seine and reaching our islands, as normally we should have expected it to have done. It is possible that similar influences may have succeeded in preventing such forms as *Emberiza hortulana*, *Acrocephalus turdoides*, and *A. aquaticus* from extending their breeding range to us.

Again, not one of the species tabulated above can be fairly regarded as dominant or widely dispersed in the British Area. The reason these species are not dominant in South-west and West-central Europe, except in one or two isolated cases which only tend to prove the rule, is probably because they were prevented by climatic and glacial conditions from reaching that area until the vast northern Emigration of birds from Refuge Area II. had taken place, which we have every reason to believe occurred earlier and under more favourable auspices—due to Gulf Stream influence—than was the case with birds whose emigrations progressed from the south-east. Before these south-eastern species had succeeded in colonizing West Europe, or before they had crossed the once glaciated portions of Central Europe, these temperate western lands were occupied by widespread and dominant species from the south which had become well established (or it may be the sea had separated Britain from continental Europe) before they arrived so far to the north-west—two powerful checks to the western progress of these eastern birds. We have every reason to believe that had these birds come up from the south, instead of from the remote south-east, they would either have been regular visitors to or residents in our area; unquestionably their most important line of North-western Emigration was to the east of Italy. We must also bear in mind our law (*conf.* p. 60) that a species in the Northern Hemisphere never normally increases its breeding range in a southerly direction—hence undoubtedly the absence of such south-eastern species not only from Iberia, and France (in some cases), but practically from the British Area, their lines of Emigration following such a course as to entail a southern—and I maintain an impossible—extension of range in order to reach such countries, from which, however, we already know they are normally entirely absent. We see by the instances tabulated above how birds can spread from the south-east even to Scandinavia, to Germany, and even to France, and yet be rare or abnormal with us and in Iberia, the dominant line of their migrations trending south-eastwards in the exact direction of their ancient north-western lines of emigration. These species have no claim whatever to be regarded as part of our avifauna; they are all of them uncommon with us, and most probably will ever remain so.

It is a most astonishing fact that there is not a single common or dominant species passing the British Islands on migration which does not either winter or breed in that area or in Refuge Area II. The great number of birds that occur in our islands sparingly or irregularly every year, neither staying to breed nor to winter, are really migrants out of their usual course, and belong to species whose dominant line of flight normally trends south-east. Comparatively speaking, a few individuals of the species in the above table, marked with an asterisk, regularly pass down the west coasts of Europe

—including the British Islands—by way of Iberia to North-west Africa to winter; but it seems very probable that the ancient Emigration of the ancestors of these individuals was originally from east to west along the Mediterranean basin, and thence north up West Europe, inasmuch as the Migration of their descendants at the present time does not extend far down the African continent (probably not beyond the limits of Refuge Area II., as these species are unknown in or only irregular stragglers to the Canary Islands) in the west, but the Migrations of the descendants of eastern individuals extend down that continent on its eastern portions even to the Cape. Moreover, the birds that winter in the west are few, and often irregular in appearance; the great bulk of the species wintering in south-eastern areas. It must always remain a moot point whether, in the course of their Post-Glacial emigration or extension northward of breeding area with the change of climate, by this route, any or all of these five species bred in the British Area when that region was of an Arctic or sub-Arctic character: probably they did so very sparingly. I may also remark ere leaving the subject that there are, of course, vast numbers of individuals of other species that breed or winter with us which pass our islands only on passage to breed further north or winter further south, but their movements do not materially affect the question we have been discussing. The significance of the above facts in relation to the past geographical conditions of Refuge Area II and of continental Africa in glacial ages cannot be too strongly impressed upon the reader.

We now come to deal with those species of which

the range base of a portion was undoubtedly in Iberia and North-west Africa, many of them with southern representatives remaining as evidence of that olden habitat, and most of which continue to visit that area in winter, or breed from North-west Africa or Iberia north to areas as high as the British Islands or even yet higher latitudes, but which, from a variety of reasons dealt with below, do not breed in our area, and can only be regarded as abnormal visitors to it.

SPECIES.	AREA NEAR BRITISH ISLES.	N. LIMITS IN EUROPE.	REMARKS.
Nucifraga caryocatactes	Black Forest; S. Scandinavia.	67	An Eastern Emigrant, with allies in Asia.
A Emberiza hortulana	N. France; Holland.	66½°	
A Galerita cristata	N. France; S. Holland.	60°	Several races in N. W. Africa.
A Anthus campestris	N. France; Holland.	57°	South African race *Anthus pyrrhonotus*
A Motacilla flava	N. France; Belgium; Holland.	60°	An Eastern Emigrant.
A Acrocephalus turdoides	N. France (Calais); Belgium; Holland.	53°	All near allies in far East.
A Acrocephalus aquaticus	N. France; Holland.	55°	
A Coracias garrulus	N. France; N. Germany; Belgium; Holland; Denmark.	61°	
Nyctea nyctea[1]	Scandinavia.	75°	
Bubo maximus[1]	France; Denmark; Germany.	71°	
A Athene noctua	France; Belgium; Holland.	56°	Allies in N. Africa and Asia.
A Scops scops ...	N. France; Belgium; Holland.	53°	Resident races in N. and S. Africa.
A Milvus ater ...	Germany.	60°	Allies in N.E. Africa and in Asia.

[1] It is probable that these species were once indigenous to the British Area, although now extinct (conf. *Ibis*, 1891, pp. 385, 386).

SPECIES.	AREA NEAR BRITISH ISLES.	N. LIMITS IN EUROPE.	REMARKS.
A Fuligula nyroca	Holland ; Germany.	55°	
A Ciconia alba ...	Holland ; Germany.	59°	Allies in E. Asia.
A Ciconia nigra ...	Holland ; Germany.	56°	Allies in Oriental and Ethiopian Regions.
A Sterna caspia ...	Holland (?) ; Island of Sylt.	59°(?)	
A Sterna anglica	Island of Sylt.	59°(?)	
A Podiceps nigricollis	Holland ; Denmark ; Germany.	56°	
A Crex parva ...	Denmark ; Holland (?) ; Central France ; Belgium (?).	56°	

Before making any comment upon the species tabulated above, it is necessary for us to deal with another group of birds, a study of whose geographical distribution and migratory movements will assist us, I think, in demonstrating the philosophy of a very intricate and hitherto inexplicable phenomenon : viz. the absence of certain species from the British Islands which are common and widely dispersed in continental areas almost within sight of them. Various ingenious suggestions have been made by naturalists in their attempts to explain what looks like an anomaly of distribution, but, as I hope presently to show, the curious fact is perfectly regular and conforms in every way to the known laws of avian dispersal. This group will be composed of species whose northern range in the *extreme west* of continental Europe, or in all Europe, does not reach the British Islands, or say latitude 50°.

RANGE BASE OR REFUGE AREA, III

Migration of British Birds Plate V

CHAPMAN & HALL LIMITED

	SPECIES.	N. LIMIT IN WEST.	N. LIMIT IN EAST.
A	Anthus spipoletta ...	Hartz Mts. 52°; Urals 64°.	Altai Mts. 50°.
	Calandrella brachydactyla	France, S. of 47½°.	S.W. Siberia 47½°.
	Accentor alpinus ...	Alps; Carpathians 49°?	Turkestan, limit unknown.
	Sylvia orphea	France, S. of 48°.	S. Russia 48° (?).
	Aëdon galactodes ...	Iberia up to 42°.	Turkestan.
A	Monticola saxatilis ...	Hartz Mts. 52°.	Lake Baikal 55°.
	Caprimulgus ruficollis[1]	S. half of Iberian Pen.	
A	Merops apiaster ...	S. of France, Alps.	Carpathians; Russia 52½° (which is as high N. as Yorkshire).
A	Cypselus melba ...	S. of lat. 50°.	Urals 55°.
A	Falco cenchris	Pyrenees 42°.	Russia 46° (an Ethiopian species).
A	Tadorna casarca ...	W. Europe 43°.	E. Europe 50°; Asia 55°.
A	Fuligula rufina	Lat. 50°.	Lat. 50°.
A	Phœnicopterus roseus	W. Europe 43°.	Asia 50°.
A	Plegadis falcinellus ...	S. France 43°; Sclavonia 46°.	Volga, S.W. Siberia 48°.
	Ardea bubulcus ...	S. half of Iberian Pen.	An Ethiopian species spread W. along N. Africa.
	Ardea comata	Iberian Pen. 42°.	S. Russia, Caspian 47°.
	Ardea garzetta	W. Europe 43°(?).	E. Europe 46°.
	Ardea alba	47° or 48°.	47° to 50° (?).
	Ardea purpurea ...	W. Europe 47°.	Siberia 55° *fide* Pallas.
A	Glareola pratincola	S. France 43°.	Siberia 46°.
A	Himantopus melanopterus	S. France 43°.	Hungary 48°; E. Russia 46°.
	Hydrochelidon hybrida	S. France 43°.	S. Russia 46°.
A	Hydrochelidon leucoptera	S. France 43°.	Poland 52°; Siberia 50°.

[1] Winter quarters unknown. I presume the line of migration to extend eastwards through Tunis and Tripoli, and thence round the desert to Central Africa. It is decidedly a species of Eastern origin.

The species contained in the two preceding tables are mostly birds that have spread northwards or north-westwards from a Glacial Refuge or Range Base in Greece, Asia Minor, Palestine, and Africa from Tripoli to Egypt in the north, south on that continent to varying limits down to the Cape Colony. They are the descendants of the individuals of the species that occupied Refuge Area III.; whilst those—if any—that inhabited the extreme west of Europe (France, Holland, British Area) were exterminated, except such as occupied Iberia and North-west Africa. By a reference to the map of Europe it will be seen how a species could extend its northern area, with the return of milder climatal conditions, even to Belgium and Holland from the eastern Refuge Area, and yet be effectually barred by the wide stretch of sea from spreading westwards to our islands, especially as ground to the north was open to the settlers. This to my mind also explains why so many of these species range into Scandinavia and Russia (the further north they go in the west the wider becoming the barrier of sea separating them from the British Islands), and are yet absent from our area. We also find that in many cases these species go much further south to winter in Africa in the east than they do in the west. We must also remember that the area of country composing the eastern Refuge Area or Range Base (III.) is vastly more extensive than that of the western Refuge Area or Range Base (II.), and reaches many degrees further north. If we follow the probable lines of Emigration after the Glacial Epoch passed away taken by these apparently abnormal species, we shall find that to reach our islands at all from the eastern Refuge Area—taking into consideration especially the glaciated condition of the Alps—*an extension of range southwards* would have had to have been made, and that is utterly opposed to known facts and to the Law already propounded. If we accept such an explanation of the phenomenon of distribution now presented by these species we can then understand why it is that species breeding commonly in Holland, Belgium, Germany, and even in some instances Scandinavia and North-central Russia, do not inhabit our area. Of the three British species that regularly migrate south-east in autumn (*conf.* p. 84) it is a most significant fact that they are common in France (directly south of the British Isles) or Central Europe, either as summer visitors or passing migrants, yet are rare or abnormal in Iberia. Of the various abnormal visitors to Britain, whose range in West Europe does not reach as high as the British Isles, no less than eight breed to the north of some part of our area east say of Ostend, and where the North Sea is 150 miles or more across; whilst others that breed a long way south of us in the west approach us much more closely in latitude in the east. Now all this appears to suggest that the species breeding so close and yet not visiting us are descended from the emigrants that re-peopled Europe after the glaciers retreated from Refuge Areas east say of E. long. 8°, spread slightly to the west of that longitude in the north in one or two instances, and in many instances attaining a higher latitude than our own in districts directly north of that Refuge Area—and present winter home. The individuals of these species breeding in Iberia or North-west Africa probably never ranged as high as our area in Pre-Glacial times; if they did, the Ice Age exterminated them, or they were entirely absent from Western Europe, which then extended down to the Canary Islands. Or did they inhabit that area, we may presume that they were sedentary as they are to-day, in the sense of not leaving it in summer or winter; or, yet again, the much smaller size of that Refuge Area (II.), and consequently fewer number of individuals, would not demand such a wide extension of northern range in Post-Glacial times, as amongst the much more numerous individuals that occupied the eastern Refuge (III.). It is even possible that the Glacial Epoch did not affect the individuals of these species in Refuge Area II. at all.

The White Stork (*Ciconia alba*), for instance, breeds commonly in Iberia and North-west Africa. It also breeds in Holland, Germany, and South Sweden, but misses the British Islands. Why? There can be little doubt that the birds breeding in Iberia were never affected by the Glacial Epoch, but the birds breeding in Holland, Germany, and Sweden undoubtedly were, and were all exterminated, especially through their inability to rear offspring. A portion of the species, however, occupied Refuge Area III., and from those Storks that peopled that area the individuals have descended that breed in North-west Europe to-day, as is to my mind surely indicated by the line of their migration at the present time, namely, across France, Italy (where they are not known to breed), and down the Danube valley, across Turkey and Asia Minor. Significantly enough, the Storks that migrate south-east from northern areas winter the furthest south in Africa, penetrating down the Nile valley to the Cape; whereas those breeding in North-west Africa and Iberia—a much more southerly area—only appear to draw south to West Africa, and in my opinion reach that locality by coasting round the Sahara east, south, and west again, as they did when that area was a sea: as is usual in such cases, the White Stork is only an abnormal and very irregular migrant to the Canary Islands.[1] Did the individuals breeding in Holland and South Sweden migrate down West Europe,—of which, however, we have no evidence,—then by a well-known law they should go to the Cape; but as they are birds breeding at the limits of the northern range, they go the furthest south, and by a route which misses South-west Europe altogether! Very similar remarks apply to the Black Stork. The Red-crested Pochard (*Fuligula rufina*) may be cited as another instance. There can be little or no doubt that the individuals of this species breeding in Europe north of Italy and east of France are the descendants of emigrants from the far east, inasmuch that the bird is very rare in the East Mediterranean and in Egypt, but can be traced through the basins of the Black and Caspian Seas to Turkestan and North Persia, thence through Afghanistan to the Refuge Area or pre-glacial Range Base in India. The individuals breeding in Spain, Sicily, Sardinia, and Southern Italy form part of the colony whose pre-glacial range extended to North-west Africa, where they are also residents.

Very interesting evidence in support of this line of Emigration is furnished by the Blue-headed Wagtail (*Motacilla flava*), and its several allied forms. The Blue-headed Wagtail is only an abnormal migrant to the British Islands, which are entirely beyond the con-

[1] There can be no doubt whatever that the Knots (*Tringa canutus*), passing our coasts on migration, reach their winter quarters in Africa by a route east along the Mediterranean and south-west across Africa to the west coasts as far as Damara Land. It is significant that this bird is unknown on the Canaries.

fines of its distribution. It is, however, one of the commonest birds in summer across the English Channel in France and Holland. It breeds in Western Europe, say from Denmark to Iberia, and probably in North-west Africa, but, as is usual in such cases in the far east, it does not breed so far south. It passes South-east Europe and North-east Africa on migration, and winters in South Africa. The Arctic form of this Wagtail (*Motacilla cinereocapilla*) is a summer visitor to North Europe (as far west as Scandinavia) and Asia (ranging from lat. 63° to lat. 68°), the European individuals passing through Central and Southern Europe and North-east Africa on migration, and wintering in Equatorial Africa; but, as if still further to indicate this dominant line of North-western Emigration, we find a colony of this race established in the Lombard Alps! Again, another race, *M. melanocephala*, is a summer visitor to Italy, Greece, Asia Minor, the Caucasus, Persia, and Turkestan, the birds breeding in Europe migrating south-east to winter in North-east Africa. Both these races are, of course, entirely unknown in Iberia and North-west Africa, and strictly abnormal wanderers to any part of West Europe say west of Denmark. The migrations of the Blue-headed Wagtail to and from North-west Europe, even to Holland and France, trend to the south-east; the migrations of the individuals breeding in Iberia, and possibly North-west Africa, are limited within that area, and do not even extend so far south as the Canary Islands. I may also remark that this Wagtail is a very late migrant to West Europe, as is usual with migrants from the south-east, appearing in Holland, etc., late in April, a month or more later than the White Wagtail, which we know winters in the south-west!

Excluding *Nyctea nyctea* and *Bubo maximus*, out of 41 species no less than 29 (marked A in the tables) conform to this rule of Emigration and Migration, and of the remaining 12, eleven do not range to as high a latitude as our islands in any part of their distribution! I may here remark that in a great number of cases I have found that western species, from Range Bases or Refuge Areas I. and II., go furthest north in the west, their range having a strong tendency to drop in the east; whilst eastern species, from Range Base or Refuge Area III., go further north in the east, their range having a similar tendency to drop in the west.

If we glance at the relative position of the British Islands to continental Europe, we shall see that wide areas of water now bar the way to all northern extensions of southern forms, and even at the narrow Strait of Dover a line of Emigration due west would have to be followed to reach our area—a direction of extension, be it remarked, rarely made across a wide water area, as a study of our avifauna and its past emigrations and present migrations abundantly proves. No part of the British area is now situated directly north of continental land without a sea-passage of nearly 20 miles separating our islands from Cape Grisnez to the South Foreland, and even in this case a route several points west of north would have to be followed. If we take a route direct north from continental land, the narrowest sea-passage varies from about 60 miles between Cape la Hague and St. Albans Head to nearly 120 miles between the north coast of Finisterre and the entrance to

Plymouth Sound. As a general rule, even narrow seas are an effectual bar to Emigration (in many cases wide rivers are equally effective barriers), and their area appears generally to have been crossed by these avian colonists whilst such seas were dry land, and previous to submergence, in such instances where we find an abundant avifauna now separated from that of adjoining districts by water areas. On the other hand, existing water areas even of comparatively wide extent are in the majority of instances no barrier to Migration or Season Flight, inasmuch as the fly-lines of migratory birds are followed with amazing persistence, and continue to be followed across recent seas, or seas slowly becoming wider as submergence progresses. There can be no doubt that at the time these routes were formed the land surface was continuous, or nearly so; indeed, in a very great number of cases the geological evidence confirms its continuity, and as birds only know, and always and invariably follow, the one route by which their area has been extended from winter quarters, refuge areas, range bases, or centres of dispersal in past ages, they must of necessity continue to follow that route, notwithstanding the gradually widening seas, or the formation of entirely new water areas, which we know has taken place during a by no means very remote past even in our own area, on that route which lies between their breeding grounds and their winter refuge.

As showing the impassable nature of a sea-barrier, and its influence on the avifauna of a sea-encircled area, I may mention that no less than 36 species of birds breed in West continental Europe within the parallels of our latitude, or in some cases much to the north of them, which are only known as rare and abnormal migrants to us, and which it may be interesting to tabulate as follows.

SPECIES.	POMERANIA.	DENMARK.	SCANDINAVIA.	N. RUSSIA.	HOLLAND.	BELGIUM.	N. FRANCE.	REMARKS.
Nucifraga caryocatactes	×		×	×				Nomadic Migrant.
Carpodacus erythrinus	×			×				Migrant.
Emberiza hortulana ...	×	×	×	×	×	×	×	Migrant.
Galerita cristata	×	×	×	×	×	×	×	Partial Migrant.
Anthus campestris ...	×	×	×	×	×	×	×	Migrant.
Motacilla flava	×	×			×	×	×	Migrant.
Lanius minor	×		×	×				Migrant.
Muscicapa parva ...	×							Migrant.
Erithacus leucocyana	×				×	×	×	Migrant.
Ruticilla tithys	×				×	×	×	Migrant.
Sylvia nisoria	×		×					Migrant.
Acrocephalus turdoides	×	×			×	×	×	Migrant.
Acrocephalus aquaticus	×	×					×	Migrant.
Hypolais hypolais ...	×	×	×		×	×	×	Migrant.
Picus tridactylus ...			×					Resident.
Picus medius	×							Resident.
Picus leuconotus ...	×							Resident.
Picus martius	×		×	×			×	Resident.
Coracias garrulus ...	×	×	×		×	×	×	Migrant.
Athene noctua	×				×	×	×	Migrant.
Scops scops	×				×	×	×	Migrant.
Bubo maximus	×	×	×	×			×	Resident.
Milvus ater	×							Migrant.
Aquila nævia	×							Migrant.
Astur palumbarius ...	×	×	×	×	×	×	×	Partial Migrant.
Tetrao bonasia	×							Resident.
Ciconia alba	×	×	×		×	×		Migrant.
Ciconia nigra	×		×					Migrant.
Scolopax major	×	×	×	×	×			Migrant.
Tringa platyrhyncha ...			×	×				Migrant.
Totanus fuscus			×	×				Migrant.
Fuligula nyroca	×	×			×	×		Partial Migrant.
Podiceps nigricollis ...	×	× ?			×			Partial Migrant.
Sterna caspia		×			×			Migrant.
Sterna anglica		×						Migrant.
Crex parva		×			×	×	×	Migrant.

More suggestive still is the evidence furnished by the Mammalia and Reptilia. As Dr. Wallace remarks

(*Island Life*, second edition, p. 339): "Germany, for example, possesses nearly 90 species of land mammalia, and even Scandinavia about 60, while Britain has only 40, and Ireland only 22. The depth of the Irish Sea being somewhat greater than that of the German Ocean, the connecting land would there probably be of small extent and of less duration, thus offering an additional barrier to migration [emigration], whence has arisen the comparative zoological poverty of Ireland." The Reptiles furnish even more significant evidence, owing to their more limited powers of dispersal. Belgium has some 22 species of reptiles and amphibia, Britain but 13, and Ireland only 4! Again, if seas were not such an important check to Emigration, we may very naturally ask why no continental species attempt to colonize the British Islands during historic time; why do none of the birds breeding so commonly almost within sight of our woods and pastures ever seek to extend their area to us? So far as I am aware there is no evidence whatever to suggest that any one species of bird has extended its range to the British Islands since their final severance from continental land at the Strait of Dover; but on the other hand we have ample proof that many species have increased their range within our limits even during historic time, as we shall learn in a future chapter (*conf.* p. 168). It follows, then, that our present avifauna was established here during land connection with Europe; not merely by the Strait of Dover, but during such time that the English Channel formed one unbroken land surface with the north of France. Even within our area we have further confirmation of this interesting fact in the comparative poorness of the avifauna of Ireland and England, especially as regards southern types; or even, yet again, the poorness of the south-west of England, in summer migrants especially, as compared with the eastern counties—a subject which also will be referred to in greater detail in a later chapter. I should say in every case the probability is that the abnormal migrants to our area of the species tabulated above are from the East. They belong to species of an eastern origin, which to reach our islands normally would either have to increase or extend their range southwards—contrary to Law—or to have crossed wide expanses of sea, which all our experience of migration and of the laws which govern the dispersal of birds over the earth's surface tend to show they are most averse to do.

In an earlier chapter I alluded to a greater extension of land area between Greenland and Europe during Post-Glacial time; it now becomes necessary, in order to render this portion of our subject tolerably complete, to enter more fully into this question and to inquire whether during post-glacial time any Palæarctic species has had a purely Nearctic origin. As may naturally be inferred by a reference to the map, Greenland—the great central land mass between Europe and America—is much too isolated from what is now continental land and too ice-clad to possess a very extensive avifauna. Avian Emigration to Greenland appears to have been as difficult from America as from Europe, even more so, the wide, deep water areas (600 miles across) of the south proving so effectual a barrier to extension of area in this direction that not a single purely American land bird (with two very doubtful exceptions) has succeeded in extending its area to the country, whilst the Wheatear is the one solitary representative of Palæarctic land birds that regularly breeds therein. On the whole, the facilities—few and meagre as they were—have been more favourable for the Emigration of Palæarctic birds to Greenland than for Nearctic species; the Faroes and Iceland indicating a once more continuous land surface between Europe and that isolated country. Equally unfavourable has this route been for the Emigration of Nearctic species eastwards into the extreme west of Europe; not a single American bird has extended its range to the Faroes, such a line of extension being contrary to the Law we have already promulgated, and but two have prolonged their emigrations beyond Greenland as far as Iceland. These are Barrow's Golden Eye (*Clangula islandica*), and the Harlequin Duck (*Fuligula histrionica*); both these species are aquatic, and both breed commonly in South Greenland. It is interesting to note that South Greenland is situated in a lower latitude than the Faroes; had the Emigration of these Ducks been from Refuge Areas I. and II., there can be no doubt whatever that they would have bred from the Faroes northwards to Iceland, just as we find some species whose Emigrations started from South-west Europe which have extended their range to and breed in Iceland, but do not breed in South Greenland, as most assuredly they would have done had their line of Emigration been from Labrador northwards. Of Nearctic origin, however, and governed by the Law which forbids an extension of breeding area southwards, both these Ducks are only abnormal visitors to any other part of Europe. There can be little doubt that had the same facilities for Emigration existed between Labrador and Greenland as between the British Isles and Iceland, the latter island would have possessed a much more heterogeneous avifauna, and that it would have been an area in which the ranges of many Palæarctic and Nearctic species would have coalesced. I cannot find that a single species (excluding the Wheatear) breeds in South Greenland and in Iceland that does not go further south in the Nearctic Region to winter—proof that some portion of the species had a range base there from which it emigrated northwards. It is a curious fact that the Wheatear is a comparatively common summer visitor to the extreme south of Greenland. The individuals that breed in Greenland, say south of lat. 65°, most certainly do not reach their summer area by way of Iceland, but must enter the country either from the south-east, south-west, or west. It is most improbable that the Wheatears breeding in South Greenland cross the Atlantic from North-west Ireland to Cape Farewell; it is even more improbable that they reach their breeding area by way of the coast of Labrador; not only because, so far as we can ascertain, the species had no base in the Nearctic region during the Glacial Epoch, but because the individuals that have been observed in Labrador are extremely few, and can only be classed as abnormal migrants there. We are therefore forced to the conclusion that the Wheatears breeding in the extreme south of Greenland are the descendants of individuals that have extended their area from Eastern Asia across Behring Sea—when dry land—into the Nearctic region. These individuals fly nearly due west across Canada and winter in Mongolia! That the Wheatear is a species of great colonizing power is proved by the enormous extent of its breeding area, which extends from the south temperate zone to the highest known limits of land.[1] The Knots that reach us from Greenland never pass south of say lat. 65° in that country, and come *viâ* Iceland to the British Islands.

Here I may remark that the fact of the Great Northern Diver (*Colymbus glacialis*) breeding in the Faroes—and possibly within the British Isles (Shetlands)—is to my mind overwhelming and positive proof that some portion, if comparatively but a few individuals, of this species dwelt in Refuge Area I. during the Glacial Epoch, and from which their descendants passed north with the return of a milder climate and open water. It may seem rather an anomalous fact that a species dwelling in Refuge Area I. should not breed in Scandinavia, but this Diver is thoroughly a marine bird, or one only able to exist on or near to open water, and its Emigrations northwards appear to have been restricted to the western portions of the British Area; at the time they were taking place the English Channel and the North Sea were dry land, and possibly still covered with ice and snow, whilst a land connection between Scotland and Iceland, by barring the progress of the warm ocean currents, kept the climate of Scandinavia and the seas adjoining too severe for the existence of any birds at all. The disappearance of the ancient isthmus or peninsula

[1] Strong evidence that the Wheatears breeding in Greenland do not reach the country *viâ* Iceland is furnished in the fact that birds have been observed in autumn to pass the extreme south of Greenland as late as the 5th of October, after having apparently been entirely absent from that district for several weeks.

between Scotland, Iceland, and perhaps Greenland, and the consequent entrance into the area beyond it of the Gulf Stream was a much later event, as I think we have some evidence to prove. Broadly speaking, the Post-Glacial Emigration of birds in West Europe from the south appears to have spread in two well-defined directions, one, which we may call Route *A*, by way of the British Area, the Faroes, and Iceland to Greenland; the other, which we will term Route *B*, by way of Holland, Belgium, and Denmark to Scandinavia. The former stream of emigration took place most probably in advance of the latter, owing to the genial influences of warm ocean currents. The birds therefore that breed or winter in the British area followed this *A* route north; the birds that followed the *B*, or continental route, north are only abnormal visitors to our shores. Of course it must be clearly understood that many species—or the individuals of many species—reached Scandinavia by an emigration across the British Area and the once dry land of the North Sea, and it is the descendants of these individuals that still continue to reach us either as passing migrants or as winter visitors from that country.

The following table contains West European species that breed in or visit the various specified island areas between the British Isles and Greenland, and which form very interesting evidence, not only suggestive of an ancient coast-line between the two countries, but of a route of past Emigration and Migration. Owing to the vast submergences which have taken place on this route, it is nothing near so important a highway for Migrants as was once the case. Sufficient evidence, however, exists to testify to its former importance.

NOTE.—One × signifies Winter Visitant; two × ×'s Breeding; M signifies Passing on Migration; M × signifies Passing on Migration and Winter Visitant; AM signifies Abnormal Migrant.

SPECIES.	ORKNEYS.	SHETLANDS.	FAROES.	ICELAND.	GREENLAND.
Turdus iliacus	×	×	× ×	× ×	
Saxicola œnanthe... ...	× ×	× ×	× ×	× ×	× ×
Motacilla alba	M	M	× ×	× ×	
Anthus pratensis	× ×	× ×	× ×	× ×	AM
Anthus obscurus	× ×	× ×	× ×		
Hirundo rustica	× ×	× ×	× ×	AM	
Chelidon urbica	× ×	× ×	× × ?	× × ?	
Plectrophenax nivalis ...	× × ?	× ×	× ×	× ×	× ×
Sturnus vulgaris	× ×	× ×	× ×	AM	AM
Corvus corax [1]	× ×	× ×	× ×	× ×	
Corvus cornix	× ×	× ×	× ×	AM	
Alauda arvensis	× ×	× ×	× ×		
Nyctea nyctea	×	×	×	×	× ×
Haliaëtus albicilla [2] ...	× ×	× ×	× ×	× ×	× ×
Falco æsalon...	× ×	× ×	× ×	× ×	AM
Phalacrocorax carbo ...	× ×	× ×	× ×	× ×	× ×
Phalacrocorax graculus	× ×	× ×	× ×	× ×	
Sula bassana...			× ×	× ×	
Anser albifrons		×	×	× ×	× ×
Anser brachyrhynchus...				× ×	
Bernicla leucopsis ...		AM	AM	× × ?	× × ?
Bernicla brenta	M	M	M	× × ?	
Cygnus musicus	M ×	M ×	M ×	× ×	× ×
Anas boschas	× ×	× ×	× ×	× ×	× ×
Anas strepera	×	×	M	× ×	
Anas acuta	×	M	× ×	× ×	AM
Anas crecca	× ×	× ×	× ×	× ×	
Anas penelope	× ×	× ×	× ×	× ×	AM
Fuligula cristata	×	×	× ×		
Fuligula marila	×	×	× ×	× ×	
Fuligula glacialis	×	× × ?	× × ?	× ×	
Fuligula nigra	× M	× M	× M	× ×	
Somateria mollissima ...	× ×	× ×	× ×	× ×	× ×
Somateria spectabilis ...	×	×	× M	× × ?	
Mergus merganser ...	× M	× M	× M	× ×	
Mergus serrator	× ×	× ×	× ×	× ×	× ×
Columba livia	× ×	× ×	× ×		
Coturnix communis ...	× ×	× ×	× ×		
Crex pratensis	× ×	× ×	× ×		AM
Rallus aquaticus	× ×	× ×	× × ?	× ×	
Fulica atra	× ×	AM	× ×	AM	AM
Ægialitis hiaticula ...	M	M	M	× ×	× ×

[1] The Ravens that breed in South Greenland are colonists from the Nearctic region, and perhaps subspecifically distinct—another indication of the route by which this area has been invaded by birds.

[2] Probably *Haliaëtus leucocephalus* in South Greenland

SPECIES.	ORKNEYS.	SHETLANDS.	FAROES.	ICELAND.	GREENLAND.
Charadrius fulvus	× ×	× ×	× ×	× ×	AM
Strepsilas interpres ...	× M	× M	× × ?	× ×	× ×
Hæmatopus ostralegus	× ×	× ×	× ×	× ×	AM
Phalaropus hyperboreus	× ×	× ×	× ×	× ×	× ×
Scolopax gallinago ...	× ×	× ×	× ×	× ×	AM
Tringa alpina	× ×	× ×	× ×	× ×	× ×
Tringa maritima	× M	× M	× ×	× ×	× ×
Tringa canutus	M	M	M	M	× ×
Tringa arenaria	M	M	M	× ×	× ×
Totanus calidris	× ×	× ×	× ×	× ×	
Limosa melanura... ...	AM	AM	× ×	× ×	
Numenius phæopus ...	× ×	× ×	× ×	× ×	AM
Sterna arctica	× ×	× ×	× ×	× ×	× ×
Pagophila eburnea ...	×	×	×	×	× ×
Larus ridibundus	× ×	× ×	× ×		
Larus canus	× ×	× ×	× ×		
Larus argentatus	× ×	× ×	× ×		
Larus fuscus	× ×	× ×	× ×		
Larus marinus	× ×	× ×	× ×	× ×	
Larus glaucus	× M	× M	× M	× ×	
Larus tridactylus	× ×	× ×	× ×	× ×	
Stercorarius catarrhactes	×	× ×	× ×	× ×	
Stercorarius richardsoni	× ×	× ×	× ×	× ×	
Alca torda	× ×	× ×	× ×	× ×	
Uria troile	× ×	× ×	× ×	× ×	
Uria grylle	× ×	× ×	× ×	× ×	
Mergulus alle	× M	× M	× M	× ×	× ×
Fratercula arctica... ...	× ×	× ×	× ×	× ×	
Colymbus glacialis ...	×	×	×	× ×	× ×
Colymbus septentrionalis	× ×	× ×	× ×	× ×	× ×
Podiceps cornutus ...	× M	× M	M	× ×	
Puffinus anglorum ...	× ×	× ×	× ×	× ×	
Procellaria pelagica ...	× ×	× ×	× ×	× ×	× ×

NOTE.—None of the species have colonized Greenland south of say lat. 65° *viâ* Iceland; those breeding in South Greenland have reached the country from Nearctic bases, or, in the case of sedentary forms, when the North Atlantic, as low as lat. 60°, was dry land. Their range extension has never taken a southern trend into this area normally. If the White-tailed Eagle is a resident in South Greenland, then its ancestors reached the country when dry land extended as low as lat. 60°. There can also be no doubt that *Ægialitis semipalmatus* and not *Ægialitis hiaticula* is the Ringed Plover breeding in the extreme south of Greenland. There is no evidence whatever to show that the Palæarctic Ringed Plover breeds south of Cumberland Bay across Davis Strait, which is not quite so far south as the southern extremity of Iceland. It is significant that Hagerup did not meet with the Semipalmated Plover. The Palæarctic Ringed Plover does not winter in any part of the Nearctic Region, whilst its Nearctic ally, though it breeds from Greenland across America to North-east Asia, winters nowhere in the Palæarctic Region —facts entirely in accord with the Law of their dispersal. I may also remark that the Lapland Buntings breeding in Greenland are from a Nearctic base—this species is unknown in Iceland.

CHAPTER IV.

THE GLACIAL RANGE CONTRACTION AND POST-GLACIAL EMIGRATION OF BRITISH BIRDS (*continued*).

Anomalous Facts—Analysis of the Facts suggested by preceding Table—Table demonstrating the two Dominant Lines of Post-Glacial Emigration in the extreme West of Europe—Analysis of Table—Variations in the Northern Limits of Species—Ancient Line of Emigration from the British Area Eastwards into Continental Europe—The North Sea Plains—Their gradual Submergence and its Effect on Birds—Table of East and North-east Emigrants—Analysis of Table—Influence of Temperature on Birds—Effects on Birds of Isolation of British Area from Continental Land—Professor Geikie on Emigration to the British Area—Emigration to Ireland—Impossibility of Southern Emigration to this Area from Scotland—Migration of Birds in the Valley of the Petchora—Emigration of Birds within the British Archipelago—Resident Species—Table of Resident Species—Analysis of Table—Absence of Birds from Ireland—Summer Migrants—Table of Summer Migrants—Analysis of Table—Table of Autumn Migrants to, and Coasting Migrants over, the British Islands—Analysis of Table—Table showing the Proportional Distribution of Species over the British Area—Deductions from the Facts—Table of Endemic British Species and Races—*Résumé* of Present and Preceding Chapters—Importance of New Law of Dispersal—Exterminating Effects of Glacial Epoch—Effects of Cold Winters—The Dartford Warbler—Importance of Southern Range Bases.

ONE or two facts embodied in the table at the close of the preceding chapter call for special mention. It may be remarked as somewhat anomalous, for instance, that the Gannet (*Sula bassana*) breeds on the Faroes and

Iceland, but does not visit the Orkneys and the Shetlands for that purpose; the Pink-footed Goose (*Anser brachyrhynchus*) breeds on Iceland, but is absent from the Faroes, the Orkneys, and the Shetlands; the Bernacle Goose (*Bernicla leucopsis*) is only an abnormal migrant to the Faroes and the Shetlands, yet probably breeds on Iceland. There can be little doubt that these species keep well out to sea in performing their annual migrations southwards, inasmuch as they are well-known visitors to the coasts of the British mainland. Again, the common occurrence of the Bean Goose (*Anser segetum*) and Bewick's Swan (*Cygnus bewicki*) off the British coasts in winter, and their absence from Iceland, suggest unknown breeding grounds due north of these areas in North-east Greenland, or in undiscovered land between Emperor William Land and Spitzbergen. Another interesting fact is presented in the migration of the Black-tailed Godwit (*Limosa melanura*) to Iceland and the Faroes. In its north-western migration this species misses the British Islands entirely, and appears to follow a route directly up the North Sea—probably the river valley which once occupied the ancient land between Britain and the continent—a vanished land which was once the breeding ground of this Godwit, as it slowly emigrated north. Godwits seem to be exceptionally attached to old routes of passage. The Bar-tailed Godwits (*Limosa rufa*) passing up our coasts in spring apparently go no further north than Spurn, and then strike out to sea as if following an ancient and submerged coast-line, or a continuation eastwards of the Humber valley (see Map, p. 21); whilst the eastern form of this Godwit (*Limosa rufa uropygialis*) appears to follow a similar sunken coast between New Caledonia and New Zealand (*Migration of Birds*, pp. 99, 100).

The significance of the facts suggested in the above table cannot be ignored by any student of Migration and Avian dispersal. Out of the 75 species enumerated, the breeding range of no less than 14 species is absolutely continuous between the two extreme areas, and these species breed in this direction from the British Islands to Greenland wherever land occurs.[1] Taking Iceland as the next most remote area on this line of dispersal, we find (including one or two doubtfuls) that no fewer than 58 species, or about eight-tenths of the whole, resort to it to breed; whilst almost exactly the same number of species (57) breed on the Faroes. In two cases at least we find that the highest northern emigration of the species throughout the world has been made along this route, which fact seems almost incredible when we bear in mind the geographical conditions of the area, and the comparative ease of a continental extension. The Gadwall (*Anas strepera*) breeds in Iceland, certainly north of lat. 64°, but only reaches lat. 57° in Scandinavia, and lat. 60° in the Stanavoi Mountains in Siberia; the Rock Dove (*Columba livia*) breeds in lat. 62° in the Faroes, but only does so up to lat. 59° in Scandinavia, its highest known northern limit elsewhere. The Chough (*Pyrrhocorax graculus*) and the Nightingale (*Erithacus luscinia*) attain their highest

[1] More information on this intricate subject is badly needed. We want particulars respecting the birds of East Greenland especially from lat. 65° northwards through Egedes Land, before we can come to any absolutely correct conclusion respecting Avian Emigration from Europe in this direction.

known northern limits in the British Islands. The migrations of the Knot (*Tringa canutus*), the Sanderling (*Tringa arenaria*), the Snow Bunting (*Plectrophenax nivalis*), and the Wheatear (*Saxicola œnanthe*), are perhaps more northerly in this direction, more dominant, stronger, than in any other part of Europe. Again, the Great Skua (*Stercorarius catarrhactes*) and the Manx Shearwater (*Puffinus anglorum*) range along this route to Iceland, and the Stormy Petrel (*Procellaria pelagica*) to Greenland, yet all three species are absent from the Scandinavian Peninsula—a fact which suggests a land or ice barrier to eastern progress beyond the longitude of the Shetlands in remote ages when the emigration north of these species was in progress. These are profoundly interesting facts. They admit of but one interpretation —a greater extension of land surface between Greenland and the British Area during earlier ages. The question of the ancient land areas in the North Atlantic scarcely comes into the subject of this work, but if the present dispersal of birds be any guide to past physical changes, there can be no doubt that a vast extension of land southwards between America and Europe once existed, and that too possibly during Tertiary time.

Our next table will indicate the two dominant lines of Post-Glacial Emigration in the extreme west of Europe —one extending to, and by way of, the British Islands to Scandinavia, principally to Norway; the other by way of Belgium, Holland, and West Germany to Denmark and Scandinavia, principally to Sweden.

NOTE.—Two × ×'s signifies Breeds; W Winter Visitor; M on Migration; WM Migration and Winter; S Straggler to; W × × Winter and Breed: SW Straggler in Winter.

SPECIES.	ENGLAND.	SCOTLAND.	ORKNEYS.	SHETLANDS.	NORWAY.	SCANDINAVIA. NORTHERN LIMITS.	SWEDEN.	DENMARK.	W. GERMANY.	HOLLAND.	BELGIUM.	SPECIES.
Turdus viscivorus	× ×	× ×				66½	× ×	× ×	× ×	× ×	× ×	Turdus viscivorus.
Turdus musicus	× ×	× ×				66½	× ×	× ×	× ×	× ×	× ×	Turdus musicus.
Turdus iliacus	WM	WM	WM	WM	× ×	71	× ×	WM	WM	WM	WM	Turdus iliacus.
Turdus pilaris	WM	WM	WM	WM	× ×	71	× ×	WM	WM	WM	WM	Turdus pilaris.
Merula merula	× ×	× ×	× ×	S		67	× ×	× ×	× ×	× ×	× ×	Merula merula.
Merula torquata	× ×	× ×	M	M	× ×	71	× ×	× ×	× ×	× ×	× ×	Merula torquata.
Saxicola œnanthe	× ×	× ×	× ×	× ×	× ×	71	× ×	× ×	× ×	× ×	× ×	Saxicola œnanthe.
Pratincola rubetra	× ×	× ×	× ×	× ×	× ×	70	× ×	× ×	× ×	× ×	× ×	Pratincola rubetra.
Pratincola rubicola	× ×	× ×	S	S				× × ?	× ×	× ×	× ×	Pratincola rubicola.
Ruticilla phœnicurus	× ×	× ×	S	S		71	× ×	× ×	× ×	× ×	× ×	Ruticilla phœnicurus.
Erithacus rubecula	× ×	× ×	× ×	S		66½	× ×	× ×	× ×	× ×	× ×	Erithacus rubecula.
Erithacus luscinia	× ×								× ×	× ×	× ×	Erithacus luscinia.
Sylvia cinerea	× ×	× ×	S	S		65	× ×	× ×	× ×	× ×	× ×	Sylvia cinerea.
Sylvia curruca	× ×	× ×				65	× ×	× ×	× ×	× ×	× ×	Sylvia curruca.
Sylvia hortensis	× ×	× ×				70	× ×	× ×	× ×	× ×	× ×	Sylvia hortensis.
Sylvia atricapilla	× ×	× ×	S	S		66	× ×	× ×	× ×	× ×	× ×	Sylvia atricapilla.
Phylloscopus sibilatrix ...	× ×	× ×					58° × ×	× ×	× ×	× ×	× ×	Phylloscopus sibilatrix.
Phylloscopus trochilus ...	× ×	× ×	S	S		70	× ×	× ×	× ×	× ×	× ×	Phylloscopus trochilus.
Phylloscopus rufus	× ×	× ×	S			66½	× ×	× ×	× ×	× ×	× ×	Phylloscopus rufus.
Acrocephalus phragmitis ...	× ×	× ×	S	S		70	× ×	× ×	× ×	× ×	× ×	Acrocephalus phragmitis.
Acrocephalus arundinaceus ...	× ×	S					58° × ×	× ×	× ×	× ×	× ×	Acrocephalus arundinaceus.
Acrocephalus palustris ...	× ×							× ×	× ×	× ×	× ×	Acrocephalus palustris.
Locustella locustella	× ×	× ×						× ×	× ×	× ×	× ×	Locustella locustella.

Accentor modularis	× ×	× ×	S			70	× ×	× ×	× ×	× ×	× ×	Accentor modularis.
Regulus cristatus	× ×	× ×	M	M	× ×	66½	× ×	× ×	× ×	× ×	× ×	Regulus cristatus.
Cinclus aquaticus	× ×	× ×	× ×						× ×	× ×	× ×	Cinclus aquaticus.
Parus palustris	× ×	× ×							× ×	× ×	× ×	Parus palustris.
Parus major	× ×	× ×	S	S		66½	× ×	× ×	× ×	× ×	× ×	Parus major.
Parus cæruleus	× ×	× ×	S	S		64	× ×	× ×	× ×	× ×	× ×	Parus cæruleus.
Parus cristatus	S	× ×				64	× ×	× ×	× ×	× ×	× ×	Parus cristatus.
Sitta cæsia	× ×							× ×	× ×	× ×	× ×	Sitta cæsia.
Troglodytes parvulus	× ×	× ×	× ×	× ×		64	× ×	× ×	× ×	× ×	× ×	Troglodytes parvulus.
Certhia familiaris	× ×	× ×	S	S		64	× ×	× ×	× ×	× ×	× ×	Certhia familiaris.
Motacilla yarrellii	× ×	× ×	× ×	M	× ×	61		× ×	× ×	× ×	× ×	Motacilla yarrellii.
Motacilla alba	few M	few M	few M	few M	× ×	71	× ×	× ×	× ×	× ×	× ×	Motacilla alba.
Motacilla sulphurea	× ×	× ×	S	S				× × 54?	× ×	× ×	× ×	Motacilla sulphurea.
Motacilla raii	× ×	× ×										Motacilla raii (France).
Anthus arboreus	× ×	× ×	S			69	× ×	× ×	× ×	× ×	× ×	Anthus arboreus.
Anthus pratensis	× ×	× ×	× ×	× ×	× ×	71	× ×	× ×	× ×	× ×	× ×	Anthus pratensis.
Anthus obscurus	× ×	× ×	× ×	× ×	× ×	71	× ×	× ×	× ×	× ×	× ×	Anthus obscurus.
Muscicapa grisola	× ×	× ×	S	S		69	× ×	× ×	× ×	× ×	× ×	Muscicapa grisola.
Muscicapa atricapilla	× ×	× ×	S			69	× ×	× ×	× ×	× ×	× ×	Muscicapa atricapilla.
Hirundo rustica	× ×	× ×	× ×	× ×	× ×	70	× ×	× ×	× ×	× ×	× ×	Hirundo rustica.
Chelidon urbica	× ×	× ×	× ×	× ×	× ×	70	× ×	× ×	× ×	× ×	× ×	Chelidon urbica.
Cotyle riparia	× ×	× ×	× ×	× ×		70	× ×	× ×	× ×	× ×	× ×	Cotyle riparia.
Pyrrhula vulgaris	× ×	× ×	S	S					× ×	× ×	× ×	Pyrrhula vulgaris.
Loxia curvirostra	× ×	× ×	W	W	× ×	66	× ×	× ×	× ×	× ×	× ×	Loxia curvirostra.
Passer domesticus	× ×	× ×	× ×	× ×		66½	× ×	× ×	× ×	× ×	× ×	Passer domesticus.
Passer montanus	× ×	× ×				67	× ×	× ×	× ×	× ×	× ×	Passer montanus.
Coccothraustes vulgaris ...	× ×	S				60	× × ?	× ×	× ×	× ×	× ×	Coccothraustes vulgaris.
Fringilla chloris	× ×	× ×	× ×	M	× ×	65	× ×	× ×	× ×	× ×	× ×	Fringilla chloris.
Fringilla cœlebs	× ×	× ×	W	W	× ×	70	× ×	× ×	× ×	× ×	× ×	Fringilla cœlebs.
Fringilla montifringilla ...	W	W	W	M	× ×	71	× ×	W	W	W	W	Fringilla montifringilla.

Species.	England.	Scotland.	Orkneys.	Shetlands.	Norway.	Scandinavia. Northern Limits.	Sweden.	Denmark.	W. Germany.	Holland.	Belgium.	Species.
Fringilla carduelis	× ×	× ×				65°	× ×	× ×	× ×	× ×	× ×	Fringilla carduelis.
Fringilla spinus	w	× ×		s		67	× ×	× ×	× ×	× ×	w ?	Fringilla spinus.
Linota cannabina	× ×	× ×	s			64	× ×	× ×	× ×	× ×	× ×	Linota cannabina.
Linota flavirostris	× ×	× ×	× ×	× ×	× ×	70		w	w	w	w	Linota flavirostris.
Linota rufescens	× ×	× ×							w	w	w	Linota rufescens.
Emberiza miliaria	× ×	× ×	× ×	× ×		58	× ×	× ×	× ×	× ×	× ×	Emberiza miliaria.
Emberiza citrinella	× ×	× ×	× ×	s		70	× ×	× ×	× ×	× ×	× ×	Emberiza citrinella.
Emberiza schœniclus	× ×	× ×	× ×	s		71	× ×	× ×	× ×	× ×	× ×	Emberiza schœniclus.
Plectrophenax nivalis	w	w × ×	w × ×?	× ×	× ×	71	× ×	w	w	w	w	Plectrophenax nivalis.
Sturnus vulgaris	× ×	× ×	× ×	× ×	× ×	69	× ×	× ×	× ×	× ×	× ×	Sturnus vulgaris.
Pyrrhocorax graculus	× ×	× ×										
Garrulus glandarius	× ×	× ×				66½	× ×	× ×	× ×	× ×	× ×	Garrulus glandarius.
Pica caudata	× ×	× ×				71	× ×	× ×	× ×	× ×	× ×	Pica caudata.
Corvus monedula	× ×	× ×	× ×	s		64	× ×	× ×	× ×	× ×	× ×	Corvus monedula.
Corvus corax	× ×	× ×	× ×	× ×	× ×	71	× ×	× ×	× ×	× ×	× ×	Corvus corax.
Corvus cornix	w	× ×	× ×	× ×	× ×	71	× ×	w	w	w	w	Corvus cornix.
Corvus corone	× ×	× ×							× ×	× ×	× ×	Corvus corone.
Corvus frugilegus	× ×	× ×	× ×	s		66½	× ×	× ×	× ×	× ×	× ×	Corvus frugilegus.
Alauda arvensis	× ×	× ×	× ×	× ×	× ×	70	× ×	× ×	× ×	× ×	× ×	Alauda arvensis.
Alauda arborea	× ×	few × ×				60	× ×	× ×	× ×	× ×	× ×	Alauda arborea.
Cypselus apus	× ×	× ×	s	s		70	× ×	× ×	× ×	× ×	× ×	Cypselus apus.
Caprimulgus europæus ...	× ×	× ×	s	s		63	× ×	× ×	× ×	× ×	× ×	Caprimulgus europæus.
Iynx torquilla	× ×	s	s	s		64	× ×	× ×	× ×	× ×	× ×	Iynx torquilla.

Picus major	× ×	s	s	s		66½	× ×	× ×	× ×	× ×	× ×	Picus major.
Picus minor	× ×	s				70	× ×	× ×	× ×	× ×	× ×	Picus minor.
Gecinus viridis	× ×	s				63	× ×	× ×	× ×	× ×	× ×	Gecinus viridis.
Coracias garrulus	s	s	s	s	·		× ×	× ×	× ×	s	s	Coracias garrulus.
Alcedo ispida	× ×	× ×							× ×	× ×	× ×	Alcedo ispida.
Cuculus canorus	× ×	× ×	× ×	× ×	× ×	70	× ×	× ×	× ×	× ×	× ×	Cuculus canorus.
Aluco flammeus	× ×	× ×					57° × ×	× ×	× ×	× ×	× ×	Aluco flammeus.
Asio otus	× ×	× ×	s	s		63	× ×	× ×	× ×	× ×	× ×	Asio otus.
Asio brachyotus	w × ×	w × ×	w × ×	w × ×	× ×	70	× ×	× ×	× ×	× ×	× ×	Asio brachyotus.
Strix aluco	× ×	× ×				64	× ×	× ×	× ×	× ×	× ×	Strix aluco.
Nyctea nyctea	W	W	W	W	× ×	71	× ×	W	W	W	W	Nyctea nyctea.
Bubo maximus	s	s	s	s	× ×	71	× ×	× ×	× ×	s	s	Bubo maximus.
Circus cineraceus	× ×							× ×	× ×	× ×	× ×	Circus cineraceus.
Circus æruginosus	M × ×	s			s	57	× ×	× ×	× ×	× ×	× ×	Circus æruginosus.
Circus cyaneus	× ×	× ×	× ×	w × ×	× ×	69	× ×	× ×	× ×	× ×	× ×	Circus cyaneus.
Buteo vulgaris	× ×	× ×	s	s	× ×	66	× ×	× ×	× ×	× ×	× ×	Buteo vulgaris.
Archibuteo lagopus	W	W	W	W	× ×	71	× ×	W	W	W	W	Archibuteo lagopus.
Aquila chrysaetus	s	× ×	× ×	s	× ×	71	× ×	× ×	× ×	× ×	× ×	Aquila chrysaetus.
Haliaëtus albicilla	W	× ×	× ×	× ×	× ×	71	× ×	× ×	× ×	W	W	Haliaëtus albicilla.
Accipiter nisus	× ×	× ×	s	s		69	× ×	× ×	× ×	× ×	× ×	Accipiter nisus.
Milvus regalis	× ×	× ×					61° × ×	× ×	× ×	× ×	× ×	Milvus regalis.
Pernis apivorus	× ×	× ×				66½	× ×	× ×	× ×	× ×	× ×	Pernis apivorus.
Falco subbuteo	× ×	s	s	s		65	× ×	× ×	× ×	× ×	× ×	Falco subbuteo.
Falco æsalon	× ×	× ×	× ×	× ×	× ×	70	× ×	× ×	× ×	× ×	× ×	Falco æsalon.
Falco peregrinus	× ×	× ×	× ×	× ×	× ×	71	× ×	× ×	× ×	× ×	× ×	Falco peregrinus
Falco tinnunculus	× ×	× ×	× ×	× ×	× ×	68	× ×	× ×	× ×	× ×	× ×	Falco tinnunculus.
Pandion haliaëtus	M	× ×	s	s		68	× ×	× ×	× ×	× ×	× ×	Pandion haliaëtus.
Phalacrocorax carbo	× ×	× ×	× ×	× ×	× ×	70	× ×	× ×	× ×	× ×	× ×	Phalacrocorax carbo.
Phalacrocorax graculus ...	× ×	× ×	× ×	× ×	× ×	70	× ×	W	W	W	W	Phalacrocorax graculus.

SPECIES.	ENGLAND.	SCOTLAND.	ORKNEYS.	SHETLANDS.	NORWAY.	SCANDINAVIA. NORTHERN LIMITS.	SWEDEN.	DENMARK.	W. GERMANY.	HOLLAND.	BELGIUM.	SPECIES.
Sula bassana	× ×	× ×	W	W			W	W	W	W	W	Sula bassana.
Ardea cinerea	× ×	× ×	S	S		68°	57° × ×	× ×	× ×	× ×	× ×	Ardea cinerea.
Botaurus stellaris ...	× ×	× ×					60° × ×	× ×	× ×	× ×	× ×	Botaurus stellaris.
Anser cinereus	W	× ×	S	S		71	× ×	× ×	× ×	× × MW	MW	Anser cinereus.
Anser albifrons	W	W	S	S	W		W	W	W	W	W	Anser albifrons.
Anser albifrons minutus ...	W					71	× ×	W	W	W	W	Anser albifrons minutus.
Anser segetum	W	W	S	S		71	× ×	W	W	W	W	Anser segetum.
Anser brachyrhynchus... ...	W	W			M			M W		M	M	Anser brachyrhynchus.
Bernicla leucopsis	W	W		S	M		M	M	W	W	W	Bernicla leucopsis.
Bernicla brenta	W	W	W	W M	M		M	W	W	W	W	Bernicla brenta.
Cygnus musicus	W	W	W	W	× ×	70	× ×	W	W	W	W	Cygnus musicus.
Cygnus bewicki	W	W	W	W				S	S	S	S	Cygnus bewicki.
Tadorna cornuta	× ×	× ×	× × ?	× × ?	× ×	70	× ×	× ×	× ×	× ×	× ×	Tadorna cornuta.
Anas boschas	× × w	× ×	× ×	× ×	× ×	$66\frac{1}{2}$	× ×	× ×	× ×	× ×	× ×	Anas boschas.
Anas strepera	× ×	× ×	× ×	× ×			57° × ×	× ×	× ×	× ×	× ×	Anas strepera.
Anas clypeata	× ×	× ×	S			$66\frac{1}{2}$	× ×	× ×	× ×	× ×	× ×	Anas clypeata.
Anas acuta	W	× ×	W	M	× ×	70	× ×	× ×	× ×	W	W	Anas acuta.
Anas crecca	× ×	× ×	× ×	× ×	× ×	70	× ×	× ×	× ×	× ×	× ×	Anas crecca.
Anas circia	× ×	S	S	S			× ×	× ×	× ×	× ×	× ×	Anas circia.

Anas penelope	w	× ×	× ×	× ×	× ×	71	× ×	× ×	× ×	× ×	w	Anas penelope.
Fuligula ferina	w × ×	w × ×	w	w	× ×	63	× ×	× ×	× ×	w	w	Fuligula ferina.
Fuligula cristata	w × ×	w × ×	w	w	× ×	66½	× ×	× ×	w × ×	w × ×	w × ×	Fuligula cristata.
Fuligula marila	w	w	w	w	× ×	70	× ×	w	w	w	w	Fuligula marila.
Clangula glaucion	w	w	w	w	× ×	70	× ×	× ×	w	w	w	Clangula glaucion.
Harelda glacialis	w	w	w	× ×? w	× ×	71	× ×	w	w	w	w	Harelda glacialis.
Somateria mollissima	× ×	× ×	× ×	× ×	× ×	71		× ×	w	w	w	Somateria mollissima.
Somateria spectabilis	s w	s w	s w	s w	s			s				Somateria spectabilis.
Fuligula nigra	× ×	w × ×	w	w	× ×	71	× ×	w	w	w	w	Fuligula nigra.
Fuligula fusca	w	w	w	w	× ×	71	× ×	w	w	w	w	Fuligula fusca.
Mergus merganser	w	w × ×	w	w	× ×	71	× ×	× ×	w	w	w	Mergus merganser.
Mergus serrator	w	× ×	× ×	× ×	× ×	71	× ×	w	w	w	w	Mergus serrator.
Columba palumbus	× ×	× ×	M	M	× ×	66	× ×	× ×	× ×	× ×	× ×	Columba palumbus.
Columba ænas	× ×	× ×	s			61	× ×	× ×	× ×	× ×	× ×	Columba ænas.
Columba livia	× ×	× ×	× ×	× ×	59 × ×							
Turtur auritus	× ×	s	s	s		65	× ×	× ×	× ×	× ×	× ×	Turtur auritus.
Tetrao tetrix	× ×	× ×				69½	× ×		× ×	× ×		Tetrao tetrix.
Tetrao urogallus	Ext. × × }	× ×				70	× ×	Ext. × × }				Tetrao urogallus.
Lagopus mutus...		× ×				70	× ×					Lagopus mutus.
Perdix cinerea	× ×	× ×	Intro. × ×			65	× ×	× ×	× ×	× ×	× ×	Perdix cinerea.
Coturnix communis	× ×	× ×	× ×	× ×		65	× ×	× ×	× ×	× ×	× ×	Coturnix communis.
Crex pratensis	× ×	× ×	× ×	× ×		66⅓	× ×	× ×	× ×	× ×	× ×	Crex pratensis.
Crex porzana	× ×	× ×	s	s		65	× ×	× ×	× ×	× ×	× ×	Crex porzana.
Crex bailloni	× ×								× ×	× ×	× × ?	Crex bailloni.
Rallus aquaticus	× ×	× ×	× ×	× ×	× ×	63	× ×	× ×	× ×	× ×	× ×	Rallus aquaticus.

SPECIES.	ENGLAND.	SCOTLAND.	ORKNEYS.	SHETLANDS.	NORWAY.	SCANDINAVIA. NORTHERN LIMITS.	SWEDEN.	DENMARK.	W. GERMANY.	HOLLAND.	BELGIUM.	SPECIES.
Gallinula chloropus	× ×	× ×	× ×	S		63°	× ×	× ×	× ×	× ×	× ×	Gallinula chloropus.
Fulica atra	× ×	× ×	× ×	S		60	× ×	× ×	× ×	× ×	× ×	Fulica atra.
Grus communis	M	M	M	M	× ×	68	× ×	M	M	M	M	Grus communis.
Otis tarda	Extinct × ×	Extinct × ×	S	S			56° × ×	× ×	× ×	S	S	Otis tarda.
Eudromias morinellus ...	M	M × ×	M × ×	M	× ×	71	× ×	M	M	M	M	Eudromias morinellus.
Ægialitis hiaticula	M	M	M	M	× ×	71	× ×	M	M	M	M	Ægialitis hiaticula.
Ægialophilus cantianus ...	× ×						× ×	× ×	× ×	× ×	× ×	Ægialophilus cantianus.
Charadrius pluvialis	× ×	× ×	× ×	× ×	× ×	71	× ×	× ×	MW	MW	MW	Charadrius pluvialis.
Vanellus cristatus	× ×	× ×	× ×	× ×	× ×	66½	× ×	× ×	× ×	× ×	× ×	Vanellus cristatus.
Strepsilas interpres	MW	M	M	M	× ×	71	× ×	× ×	M	MW	MW	Strepsilas interpres.
Hæmatopus ostralegus ...	× ×	× ×	× ×	× ×	× ×	71	× ×	× ×	× ×	× ×	× ×	Hæmatopus ostralegus.
Recurvirostra avocetta ...	Ext. × ×	S	S	S				× ×	× ×	× ×	× ×	Recurvirostra avocetta.
Phalaropus hyperboreus ...	S	× ×	× ×	× ×		71	× ×					Phalaropus hyperboreus.
Scolopax rusticola	× ×	× ×	× ×	× ×	× ×	66½	× ×	× ×	× ×	× × ?	× × ?	Scolopax rusticola.
Scolopax gallinago	× ×	× ×	× ×	× ×	× ×	70	× ×	× × ?	× × ?	× × ?	W	Scolopax gallinago.
Scolopax gallinula	W	W	W	W	× ×	71	× ×	W	W	W	W	Scolopax gallinula.
Tringa alpina	W × ×	W × ×	W × ×	W × ×	× ×	71	× ×	× ×	× ×	W	W	Tringa alpina.
Tringa maritima	W	W	M	M	× ×	71	× ×	W	W	W	W	Tringa maritima.
Tringa canutus	MW	MW	M	M	M			M	M	M	M	Tringa canutus.
Calidris arenaria	M	M	M	M	M			M	M	M	M	Calidris arenaria.
Machetes pugnax	M × ×	M	M	M	× ×	71	× ×	× ×	× ×	× ×	× ×	Machetes pugnax.

Totanus hypoleucus	× ×	× ×	× ×	× ×	× ×	71	× ×	× ×	× ×	× ×	× ×	Totanus hypoleucus.
Totanus glareola	× ×	S				71	× ×	× ×	× ×	× ×	× ×	Totanus glareola.
Totanus ochropus	M	S				67	× ×	M	M	M	M	Totanus ochropus.
Totanus calidris	× ×	× ×	× ×	× ×	× ×	70	× ×	× ×	× ×	× ×	× ×	Totanus calidris.
Totanus fuscus	M	S	S			70	× ×	M	M	M	M	Totanus fuscus.
Totanus glottis	M	× ×	S	S		70	× ×	M	M	M	M	Totanus glottis.
Limosa melanura	M	S	S	S		66½	× ×	× ×	× ×	× ×	× ×	Limosa melanura.
Numenius arquata	× ×	× ×	× ×	× ×	× ×	71	× ×	× ×	× ×	× ×	× ×	Numenius arquata.
Numenius phæopus	M Ext.	M	× ×	× ×	× ×	71	× ×	M	M	M	M	Numenius phæopus.
Hydrochelidon nigra	× ×	S	S	S		60	× ×	× ×	× ×	× ×	× ×	Hydrochelidon nigra.
Sterna anglica	S							× ×	S	M	M	Sterna anglica.
Sterna cantiaca	× ×	× ×						× ×	× ×	× ×	× ×	Sterna cantiaca.
Sterna caspia	S							× ×	M	× ×?	M	Sterna caspia.
Sterna hirundo...	× ×	× ×				66½	× ×	× ×	× ×	× ×	× ×	Sterna hirundo.
Sterna arctica	× ×	× ×	× ×	× ×	× ×	71	× ×	× ×?	M	M	M	Sterna arctica.
Sterna minuta	× ×	× ×	× ×					× ×	× ×	× ×	× ×	Sterna minuta.
Larus ridibundus	× ×	× ×	× ×	× ×		59	× ×	× ×	× ×	× ×	× ×	Larus ridibundus.
Larus canus	W	× ×	× ×	× ×	× ×	71	× ×	W	W	W	W	Larus canus.
Larus argentatus	× ×	× ×	× ×	× ×	× ×	71	× ×	× ×	W ?	W ?	W ?	Larus argentatus.
Larus fuscus	× ×	× ×	× ×	× ×	× ×	71	× ×	W	W	W	W	Larus fuscus.
Larus marinus	× ×	× ×	× ×	× ×	× ×	71	× ×	W	W	W	W	Larus marinus.
Larus glaucus	W	W	M	M	× ×	71	W	W	W	W	W	Larus glaucus.
Larus tridactylus	× ×	× ×	× ×	× ×	× ×	71	W	W	W	W	W	Larus tridactylus.
Stercorarius catarrhactes ...	W	W	W	× ×				W	W	W	W	Stercorarius catarrhactes.
Stercorarius richardsoni ...	M	× ×	× ×	× ×	× ×	71	× ×	M	M	M	M	Stercorarius richardsoni.
Stercorarius buffoni	M	M	M	M	× ×	71	× ×	M	M	M	M	Stercorarius buffoni.
Alca torda	× ×	× ×	× ×	× ×	× ×	69	× ×	M	M	M	M	Alca torda.
Mergulus alle	W	W	W	W	W			W	W	W	W	Mergulus alle.
Uria troile	× ×	× ×	× ×	× ×	× ×	71	× ×	× ×	W	W	W	Uria troile.
Uria grylle	W	× ×	× ×	× ×	× ×	71	× ×	× ×	W	W	W	Uria grylle.
Fratercula arctica	× ×	× ×	× ×	× ×	× ×	71	S	M	M	M	M	Fratercula arctica.
Colymbus arcticus	W	× ×	× ×?			71	× ×	W	× ×	W	W	Colymbus arcticus.

SPECIES.	ENGLAND.	SCOTLAND.	ORKNEYS.	SHETLANDS.	NORWAY.	SCANDINAVIA. NORTHERN LIMITS.	SWEDEN.	DENMARK.	W. GERMANY.	HOLLAND.	BELGIUM.	SPECIES.
Colymbus septentrionalis ...	W	× × few	× ×	× ×	× ×	71°	× × 57°	W	W	W	W	Colymbus septentrionalis.
Podiceps cristatus	× ×	× ×					× ×	× ×	× ×	× ×	× ×	Podiceps cristatus.
Podiceps rubricollis	W	W	S	S		66½	× ×	× ×	× ×	× ×	W	Podiceps rubr.collis.
Podiceps nigricollis	S	S	S	S				× ×?	× ×	× ×	× ×	Podiceps nigricollis.
Podiceps cornutus	W	W	W	W	× ×	68	× ×	× ×	W	W	W	Podiceps cornutus.
Podiceps minor	× ×	× ×	× ×	W	× ×	62	× ×	× ×	× ×	× ×	× ×	Podiceps minor.
Puffinus anglorum	× ×	× ×	× ×	× ×	S W			W	W	W	W	Puffinus anglorum.
Procellaria pelagica	× ×	× ×	× ×	× ×				W	W	W	W	Procellaria pelagica.
Procellaria leachi	W	× ×						W	W	W	W	Procellaria leachi.

NOTE.—Two × ×'s signifies breeds; W Winter Visitor; M on Migration; WM Migration and Winter; S Straggler to; W × × Winter and Breed; SW Straggler in Winter.

The above table brings to light some profoundly interesting facts. Taking France and Germany, say, as our base, the symbols attached to the various birds on the left side of the table indicate the emigrations of these species in the British Islands, or entirely across that area to Norway; whilst the symbols attached to the same species on the right side of the table indicate the emigrations of other individuals of the same species across Belgium and Holland to Denmark, Sweden, and Norway in that direction. These latter symbols may also indicate emigrations from more southern areas east say of E. long. 10°. The table contains 211 species. Ninety of these have unquestionably reached Scandinavia by Emigration across the British area, as well as by way of Belgium, Holland, and Denmark, as is proved by the present lines of Migration followed by those species, but in a few cases the palpable scarcity of the migrants across our area shows that the dominant line of northward extension was continental. The White Wagtail, the Black-tailed Godwit, the Crane, and the Sclavonian Grebe may be cited as instances that confirm this. The northern Emigration of a very high percentage of individuals, however, was continental. No less than 85 species have undoubtedly reached Scandinavia by way of Belgium, Holland, and Denmark only—a fact which is proved by the absence of those species from the northern portions of our area, or their northern extension on both routes not being sufficient to render it by any possibility continuous at its highest limits. Continuing the analysis still further, we find that only three species have succeeded in reaching Scandinavia—or rather Norway only—by an emigration across our area, a fact which is proved by the absence of such species from

Sweden. These are the Twite, the Pied Wagtail, and the Rock Dove. We also find that the emigrations of twelve species have only extended as far north as Denmark; whilst five have only reached the southern shores of the Baltic. Of these latter, however, it is interesting to note that no less than four are represented in the more northern latitudes by closely allied forms from which it is very doubtful whether they are specifically distinct, except in one instance. These are the Nightingale, represented in Scandinavia and Denmark by *Erithacus philomela*, the Dipper in Scandinavia by *Cinclus melanogaster*, the Bullfinch in Scandinavia by *Pyrrhula major*, and the Carrion Crow by *Corvus cornix*. Two of these birds are practically sedentary, their migrations, if any, being limited; the Nightingale is a migrant, but its line of passage is south-east, which almost completely isolates it from the western race *E. luscinia*. Of the twelve species that only reach Denmark, one is represented further north by a local race, viz. the Nuthatch, whose Scandinavian allied form is *Sitta europæa*. Again, but ten species have extended their emigrations across the British Area to a higher limit than they have done in West Continental Europe. They are the Nightingale, the Gray Wagtail, the Lesser Redpole, the Chough, the Carrion Crow, the Lesser Tern, the Great Skua, and the three species of Petrels—all, save one (the Carrion Crow), it may be remarked, thoroughly western types. On the other hand, no less than 74 species range lower in our area than in West Continental Europe. It is to my mind an astonishing fact of distribution—taking into consideration the mild climatal conditions of our isolated area, encircled as it is by the warm waters of the Gulf Stream—that so vast a percentage of species should range higher in continental Europe than they do in the British Islands. In many cases the difference of northern limit in the two areas is enormous; and curiously enough some of the most striking examples are to be found amongst such delicate species as the Thrushes and Warblers! Our Missel Thrush and Song Thrush are by no means dominant birds in Scotland, yet in Scandinavia they range up to the Arctic Circle; the Redstart is only locally distributed in Scotland, but in summer it ranges in Norway to the North Cape, the Arctic land of the midnight sun! The two species of Whitethroat are decidedly rare and local north of the Border, yet in Scandinavia they are regular summer visitors up to lat. 64°; whilst the Garden Warbler, local even in the south of Scotland, prolongs its migrations to the North Cape in Norway. The Sedge Warbler,[1] too, goes as far north in summer, yet does not range as high in our area to breed as the Orkneys. The Tree Pipit is rare north of the Clyde, yet is a regular summer visitor to the extreme north of Norway, fifteen degrees of latitude higher. Even the delicate fastidious Wood Lark—a typical English species—ranges up to lat. 60° in Scandinavia. The Swift does not breed in the Orkneys, yet does so in Scandinavia in lat. 70°; the Wryneck, unknown in Scot-

[1] The Sedge Warbler is said to be absent from the south of Norway, which is strong evidence that the species entered that portion of Europe from the south-east, and not by a northern emigration or range expansion either up what is now west continental Europe, or across the British Area. From such a South-eastern base an extension of range southwards or across a wide water area would be necessary for the species to enter south Norway—a line of emigration contrary to the law of its dispersal. Very similar remarks apply to the Wood Wren, which is absent from Norway altogether, yet breeds in Sweden as far north as Upsala.

land, goes up West Europe to lat. 64°. The three Woodpeckers only known to breed in England in the British Area penetrate in Scandinavia to lat. 63° (*Gecinus viridis*), to the Arctic Circle (*Picus major*), and to lat. 70° (*Picus minor*)! The Kentish Plover only breeds on the south coast of England in our area, but on the Continent it visits the south of Sweden for that purpose. The Ruff formerly bred in England—there is no evidence to suggest that it ever bred further north in our islands—yet it goes in summer as far north as continental land extends in West Europe. Even such hardy aquatic species as the Black-throated Diver and the Common Tern breed no further north in our area than the Scottish mainland, perhaps in the Orkneys, yet the former is found in Scandinavia to the highest limits, whilst the latter reaches the Arctic Circle.

What is the explanation of these apparently anomalous facts? We have already seen that birds are loth to extend their emigrations across water areas, and we might reasonably assume that our isolated position was a check to any considerable increase of northern range in that direction; but I am convinced that these wonderful variations in the northern limits of so many species are not due to such a cause. After a prolonged and careful study of the facts, I consider that this discrepancy of distribution is entirely due to the dominant line of Emigration followed by these species. The range base of the vast bulk of the individuals of these species breeding east say of E. long. 10° was in the south-east during the Glacial Epoch. Their normal line of Post-Glacial Emigration north and north-west from that Range Base or Refuge Area (III.) would therefore be entirely beyond our limits. The i dividuals breeding in our islands refuged in the south-west (II.) were fewer in number—as has previously been suggested—and they winter at their ancient base where they refuged; the ancestors of the individuals of the same species that range so high in Scandinavia had their Range Base in the south-east, and they migrate in autumn south-east towards that ancient refuge area, none of them normally visiting the British Islands or South-west Europe at all. Fortunately one or two species prove, even to demonstration, that a vast emigration of birds did take place in this direction. The Blue-throated Warbler (*Erithacus suecica*) breeds in Scandinavia, passes Central and South-eastern Europe on migration, and winters in North-east Africa, south to Abyssinia. The Eastern Nightingale (*Erithacus philomela*) breeds in Scandinavia and Denmark, passes Central Europe on migration, and winters in North-east Africa. The Black-throated Diver (*Colymbus arcticus*) breeds in Scandinavia and the Baltic Provinces, crosses Europe in a south-easterly direction on passage to its winter quarters in the Black Sea, occasionally wandering south to the Italian lakes, the Adriatic, and the Mediterranean. I may here remark that Mr. George Lindesay, in his very interesting paper, "Rambles in Norsk Finmarken" (*Fortnightly Review*, November 1894, p. 674), observed and recorded this strongly marked direction of the migration of birds in spring to that area. He says: "Like most of the true birds of passage, these small birds reach their northern breeding-ground from the East, coming by way of Russia and the Baltic Provinces."

Many of the species tabulated above have ceased to breed in our islands, some only in the extreme north of them, but they unerringly indicate the line of their Emigration across this area in earlier ages, when they undoubtedly did so, by visiting us on passage or remaining with us during the winter. There is also some indication—if now slight—of birds passing along some of our southern coasts (where the Channel is narrowest) on their way north to Denmark and Scandinavia—descendants probably of birds whose emigrations extended up the Channel when that area was an ancient land surface, a great river valley.

Before leaving this portion of the subject, it is necessary to make a brief allusion to an ancient line of Emigration from the British Islands eastwards into continental areas. We have already dwelt upon the probable condition of the North Sea Area during early Post-Glacial time. That this area at no very remote time, comparatively speaking, was a broad plain watered by a central river flowing into the Arctic Ocean between the Shetlands and Norway, and receiving as its tributaries not only our own eastern rivers but those of West Europe north of Belgium, there can be no reasonable doubt whatever. This plain was probably well wooded, studded with lakes and swamps, and in every way suited to the requirements of great numbers of birds, especially the earliest colonists from the south after the third glacial period had passed away. We have also every reason to believe that, owing to the influence of the Gulf Stream, West France, the English Channel (then dry land), and the British Area were able to support arboreal species of birds long before the North Sea Plains, Belgium, Holland, West Germany, Denmark, and Scandinavia were in such a condition, owing it may be to the land connection between Scotland and Iceland keeping out the warm currents, and the English Channel and North Sea being then a continuous land mass.[1] With the return of the more genial climate birds soon began to emigrate from the southern Refuge Areas (I. and II.), following the retreating ice and snow-fields north; so that by the time the climate of the North Sea Plains and West Continental Europe was sufficiently genial, a dominant resident avifauna chiefly composed of hardy species was already established in our area. With the growth of vegetation on this prehistoric plain and in West Europe the birds began to extend (or in other words to emigrate) their range east across that area. The climate, however, was too severe in winter to allow of these birds becoming resident; they merely migrated in spring further and further east each century to breed, coming back in autumn to winter in the mild climate of the Gulf Stream laved west. Conditions favourable to successful colonization and increase in Europe continued, and these West to East migrants multiplied accordingly spreading east and north-east across Europe even to West Asia, yet compelled by the severer climate of the north and east to return west every autumn. But gradually the sea began to encroach upon the land; the plains between the Continent and our eastern borders slowly became more and more water-logged as submergence went on, and the great central river valley became perhaps a fjord, which, however, the eastern migrants found no difficulty in crossing. Slowly each century this ocean inlet became wider; the water

[1] We must also not overlook the fact that the fourth Glacial Period—the epoch of the Great Baltic Glacier—which did not affect our area or the continental lands adjoining it to any great extent, must have had a vast influence on the emigration of species.

passage more prolonged, but the migratory birds with unyielding persistence continued to cross and re-cross it; until in the course of ages the North Sea occupied the land, and the contour of our islands and the opposite continental coasts gradually became as it exists to-day. Birds continue to migrate in countless hosts across this wide sea-passage, their ancestors having done so in the remote past when dry land replaced the sea; and no single generation of birds has been able to notice any portion of the vast change which centuries of submergence has accomplished. This is a sufficient explanation of the wonderful migration which takes place in spring and still more marked in late autumn across the North Sea to our islands now. The principal species that have emigrated in the past by this route, and which continue to migrate along it in the present, are specified below.

Species whose lines of Emigration from our area extend East alone.	Species whose lines of Emigration from our area extend both East and North-East
Turdus viscivorus.	Merula merula.
Turdus musicus.	Erithacus rubecula.
Pratinola rubicola.	Regulus cristatus.
Accentor modularis.	Fringilla chloris.
Parus cæruleus.	Fringilla cœlebs.
Parus major.	Sturnus vulgaris.
Troglodytes parvulus.	Corvus monedula.
Fringilla carduelis.	Corvus cornix.
Linota cannabina.	Alauda arvensis.
Passer montanus.	Asio brachyotus.
Emberiza miliaria.	Columba palumbus.
Emberiza citrinella.	Vanellus cristatus.
Emberiza schœniclus.	Grus communis.
Garrulus glandarius.	Cygnus olor.
Corvus corone.	
Corvus frugilegus.	
Accipiter nisus.	
Otis tarda.	
Podiceps rubricollis.	
Botaurus stellaris.	

Of the species whose lines of Emigration from the British Islands extend East alone, not one, it may be remarked, has succeeded in extending its breeding range across our area to Scandinavia. They are all species whose northern breeding range in the British Isles does not reach the Shetlands, or in the few instances in which it does reach that locality, the evidence distinctly shows that they have only done so within recent time. It is therefore more than doubtful whether any individuals of these typical Eastern migrants come from continental areas north or east of the Baltic, or south of Holland. From how far north in Russia these eastern migrants may come it is at present impossible even to conjecture. That the line of Emigration, however, extended into Western Asia, say as far south as Orenburg, is to some extent proved by the odd individuals of thoroughly eastern or Asiatic species that from time to time get into this western stream of migration and abnormally turn up on Heligoland, in the British Islands, and elsewhere. It may also be remarked that few, if any, of these typical Eastern migrants are dominant during winter in the extreme south of Europe, still less so across the Mediterranean, where some of them are replaced by nearly allied species or races.

Of the species whose lines of Emigration from the British Islands extend both East and North-east, it is significant that in every case they are birds that have reached Scandinavia by both routes, as is proved by their breeding dominantly in the Shetlands, and by their passing our entire eastern area on migration. It is, however, worthy of remark that in many, if not in all, cases the North-east migrants are the first to visit us, the Eastern migrants being later in their arrival. Were it only possible to separate the individuals, I think we should find that so far as the typical Eastern migrants are concerned, they all journey practically about the same time, and come from the same areas. I think this phenomenon of an East to West migration in autumn will also explain why such species as the Crane (*Grus communis*), the Bittern (*Botaurus stellaris*), and the Great Bustard (*Otis tarda*), are now only winter visitors to our islands. All these species a century or so ago were common summer visitors to the British Islands, the individuals breeding in our area migrating south in autumn to their accustomed winter quarters; but on the other hand, our islands were visited in winter by other individuals of these three species that were returning to their old quarters, from whence their ancestors emigrated East in past ages, and which they continue even to the present day to regard as their winter home. Several other instances confirming the above remarks might also be given.

Singularly enough, the individuals breeding in the British Islands of many of the species (perhaps of all) tabulated above migrate south in autumn, their places being taken by these migrant individuals of the same species from the east. This seems to be a strangely anomalous fact, yet it is one that shows how very complicated the whole subject of Avian Migration undoubtedly is. As I said in the *Migration of Birds* (p. 248), I am strongly disposed to think that Temperature has a good deal to do with this very complex movement. Besides, there can be no doubt that the individuals of these species that breed in Britain are later arrivals in our area from the south, later emigrants, more delicate of constitution, and requiring a higher winter temperature than the descendants of the earlier emigrants of these self-same species that reached us, it may be, thousands of years before, when the climate was much more rigorous, and which, as we have seen, extended their breeding range into the colder regions of the North and East, after having become thoroughly acclimatized to our winter temperature by long residence therein.

I may also remark that no birds wintering exclusively south of our area follow this route, a fact which seems to prove that our islands were a base of ancient Eastern Emigration from which resident individuals alone extended their area east. By the time most of our Summer Migrants had reached us, the North Sea had probably established a fatal barrier to all Eastern Emigration of southern forms in this direction. Not a single Warbler, in fact not a single Passerine or Picarian species that now visits us only in summer to breed, is by any strange chance observed normally to follow this East to West migration in autumn, or West to East return passage in spring.

Want of space and time, and the sad deficiency of more exhaustive information, I regret to say, prevent me from entering into greater detail concerning this particular line of Flight, which I may add possesses a singular fascination for me. The subject is too vast and too complicated to be exhaustively treated in the present volume, and in the existing state of our knowledge, but it is a portion of the science of Avian Season Flight that will have to be faced and thoroughly exhausted in the near future.

We will now proceed to analyze a little more closely the distribution of birds within the limits of the British Archipelago, and to ascertain whether the same laws, the same general conditions of dispersal, have similarly exerted their influence on the range of species in our area, as we have already learned they have governed the distribution of birds elsewhere. We have already seen that the ultimate isolation of the British Islands from continental land has had a marked effect upon the distribution of terrestrial animals; and in like manner the isolation of Ireland from the greater land mass of England and Scotland is significantly indicated by a similar disparity in its mammalian and reptilian fauna. I have in the present volume repeatedly insisted upon the aversion displayed by birds to extend their emigrations, or to increase their range across wide water areas. Instances innumerable in all parts of the world might be given in support of this assertion, but for the present purpose we will confine our confirmatory evidence to the species composing our own avifauna alone.

As I have already attempted to show, the land mass of Ireland became isolated from England before England itself became detached from continental land. There is some evidence to suggest that the north-east of Ireland continued to be joined to South-west Scotland for a much longer period, but as I hope presently to demonstrate, this connection was of no service in assisting the emigration of birds or animals to the Irish area, although it may, and probably did, help them on their emigrations northwards. The Law propounded at the beginning of the previous chapter emphatically forbids such a line of extension. Professor Geikie (*Prehistoric Europe*, pp. 511-513) very rightly remarks that "the Scandinavian type in the British Isles, as is well known, attains its greatest development in the Scottish Highlands. It is less well represented in the southern uplands of Scotland, the hilly districts of Cumberland, and the Welsh mountains, while Ireland shows a very meagre assemblage of alpine and sub-alpine forms. The Germanic type, on the other hand, is everywhere present, overspreading the other floras and giving a general character to the vegetation." As the late Mr. Forbes wrote: "Its scarcer forms are of much interest, from the clear manner in which they mark the progress of the flora and the line it took in its advance westwards. Thus we find a number of species which are still limited to the eastern counties of England, while others, which have extended over considerable tracts or into several districts of England or Scotland, have not found their way to Ireland. It is remarkable that certain species of this flora which flourish best on limestone . . . are not found in the limestone districts of Ireland, and in like manner certain species which everywhere, when found, delight in sand . . . are also wanting in such Irish localities as are best adapted for them. The fauna which accompanies this flora presents the same peculiarities, and diminishes towards the north and west. This is very observable both among the native vertebrate and invertebrate animals. Thus, among quadrupeds the mole, the squirrel, the dormouse, the polecat, and the hare (*Lepus timidus*) are confined to the English side of St. Georges Channel, not to mention smaller quadrupeds. So it is also with the birds of short flight; so most remarkably, no less than half the species being different, with the reptiles; so also with the insects and the pulmoniferous mollusca." Professor Geikie, in criticizing Forbes's remarks, continues: "These peculiarities of distribution Forbes has accounted for by supposing that Ireland was separated from England by the influx of the Irish Sea before the species, less speedy of diffusion, could make their way into the sister island, and this view has been repeated by every writer who has touched upon the question since the appearance of Forbes's famous essay. But a glance at the Admiralty's chart of the Irish Sea shows us that there is no necessity for inferring that the arrestment of the migration [emigration] was due to submergence. Were the whole British area to be elevated for six hundred feet or thereabout, the Irish Sea would disappear, but Ireland would still be separated from England by a great and deep lake, averaging twenty-five miles at least in breadth, and extending from what is now the Sound of Jura in Scotland down through the basin of the Irish Sea to a point between Braich-y-pwll in Carnarvon and Greenore Point in Wexford. This lake receiving the tribute of many Scottish, Irish, and English streams, would discharge a broad river from its lower end, which might well be impassable by many of the smaller vertebrates. That it was rather the presence of this lake and the obstacle of the Welsh mountains than the premature appearance of the Irish Sea which arrested the westward migration [emigration] of plants and animals, is shown by the remarkable fact pointed out by Professor Leith Adams that the mammalian fauna of Ireland agrees more closely with that of Scotland than of England; while Dr. Buchanan White has shown that Ireland has probably derived some of its alpine lepidoptera from Scotland. We may suppose that the temperate mammals gained admittance to Ireland from the west of Scotland, between which and the north of Ireland there was a broad land connection. Some of the larger mammals, however, such as the great Irish deer (*Cervus megaceros*), may quite well have entered Ireland from the south, crossing the river that flowed south through St. Georges Channel. But it may be questioned whether the reindeer immigrated by the same route. So far as the geological evidence goes, we have no reason to believe that at the commencement of the Post-Glacial period the British area was much more extensive than it is at present. The sea was then retiring, as we know, from the low grounds of Scandinavia and Scotland, and from the borders of East Anglia, and thus the probabilities are that when the Scandinavian flora had commenced its northward advance St. Georges Channel still separated England and Ireland. This being so, the reindeer could not at that time reach the latter country. By and by, however, the Irish Sea gradually disappeared, and a land connection took place between Scotland and Ireland, across which the alpine and sub-alpine flora and the reindeer would migrate [emigrate]. It is perhaps owing to the late appearance of this land connection that the Scandinavian type of vegetation is so poorly represented in the Hibernian flora. The climate we may suppose was already becoming milder, and the high alpine forms were gradually vanishing from the low grounds, so that only a few of these could make their way south into Ireland."

The first part of Professor Geikie's remarks very forcibly suggests that early Post-Glacial Ireland formed one land mass in the south with England, and thus presented no barrier to the emigration to that area of the various alpine and sub-alpine floræ (and of course the hardiest species of birds), relics of which floræ exist there down to the present time. But as the climate moderated, this flora, which sought and now occupies a congenial habitat on the Welsh, English, and Scotch mountains, would to a very great extent disappear from Ireland, owing to the absence of such suitable mountain haunts; and this flora was, as we know, (and still is) replaced by the dominant and southern Germanic types which must also have entered Ireland *before* the connecting land masses in the south disappeared. The fact insisted upon by Professor Geikie in support of some of his views, that the mammalian fauna of Ireland agrees more closely with that of Scotland than with that of England, is due entirely to the smaller amount of competition to which that fauna has been exposed by the invading Germanic or southern types which have established themselves in England so dominantly, together with the later appearance of that fauna in England, when the difficulties surrounding emigration to the Irish area were rapidly increasing, or perhaps almost insurmountable, due to progressing submergence and widening of the water ways, as well as to the similarity of climate and general conditions between Scotland and Ireland. Singularly enough, Professor Geikie shortly afterwards states that so far as the floras are concerned, that of Hibernia is very poor in Scandinavian—and by inference Scotch—types. Again, Professor Geikie supposes that the temperate mammals as well as many alpine forms of plants, etc., reached Ireland *viâ* the west of Scotland, when a broad land connection existed between the two areas,—but what evidence, I ask, is presented by birds, creatures even of flight to which a water barrier seems ineffectual, that any such line of extension was followed? None, absolutely none whatever! As we have already had abundant and convincing proof, and hope still to have more; the emigration of birds to our area has been from the south, south-east, or east; not a single species has entered it during Post-Glacial time from the north, nor increased its range across it from that direction. One is astounded to read of Professor Geikie insisting upon a descent into Ireland of alpine forms from Scotland to account for their scanty presence in that country. Rather must we look upon these types as relics, not as the result of abortive and abnormal emigration, but of the dominant flora that in early Post-Glacial time held the land until a changing climate destroyed its predominance: relics I may say that to my mind beyond all doubt demonstrate that St. Georges Channel and the English Channel were dry land when the Post-Glacial Emigration north of the earliest flora and fauna took place. It is impossible to have a southward extension of range progressing concurrently with an amelioration of climate. The very fact that species are advancing north with favourable conditions north of them, and, of course, in many cases, by inference, less favourable conditions south of them, is a fatal objection to any such line of dispersal. Species would not range north of any area from which they were absent until conditions in that area were less suitable, and therefore impossible for southern extension towards it. The very conditions that are driving species north, or attracting an emigration movement, are in like manner prevailing in other areas immediately east and west, and thus rendering impossible a southern emigration, or what is better described as a retrograde movement.

One of the most intensely interesting facts bearing upon this Law of Northern Extension is that of the migration of birds in the valley of the Petchora. Messrs. Seebohm and Harvie Brown, whilst stationed at Ust Zylma in the spring of 1875, kept careful observation on the migration of birds north along this river to the North Russian tundras. They report (*Rowley's Orn. Miscell.*, 1876, i. p. 245) that "there can be no doubt that Ust Zylma lies somewhat out of the line of migration." These gentlemen give a list of no less than 13 species of birds that were all common summer migrants to the tundra, yet did not pass this station on the great river. The reason for this is obvious. In the first place, we must keep in mind the fact that the present line of Migration of a species follows the past line of Emigration or range extension of that species, and that this emigration never takes place in the Northern Hemisphere in a *southerly* direction (*conf.* p. 60). If we refer to the map, we find that the Petchora, in very nearly north latitude 66°, takes a sudden trend to the south-west for nearly 150 miles, amounting in the aggregate to as much as 50 miles south of north. Birds, then, in emigrating into North Russia from the south-east, and following the river valleys, would therefore have been under the necessity of breaking the Law of their dispersal by increasing their area south-west had they kept to the river; but following the normal conditions of their dispersal, they left the valley entirely at the apex of this southern trend and began to colonize the tundras northwards. To this day their descendants miss the river valley probably entirely, after reaching its most northerly trend! These birds were all emigrants from South-west Asia or South-east Europe. On the other hand, the great body of migrants to the tundras did pass through Ust Zylma, but they either did not follow river valleys so closely (as for instance in the case of the Siberian Chiffchaff), or most certainly came from more south-westerly areas (*conf.* footnote, p. 127).

We will now endeavour to trace more minutely the emigration of birds within the British Archipelago. We will deal first with the species that are resident in the British Islands. For the most part these consist of hardy birds, numbering amongst them some of the earliest to reach our area after the third cold period of the Glacial Epoch. From these early arrivals, as we should naturally expect to be the case, the dominant avifauna of Ireland has descended. They represent species whose emigrations to the British Area took place whilst that area was much more compact than it is now; when St. Georges Channel and the English Channel did not exist, and extension northwards and westwards was not retarded or absolutely prevented by wide areas of water. Our resident birds are tabulated below.

NOTE.—× signifies Breeding; VL Breeding, but very locally; L ditto, but locally; W Winter only; VLW Very rare and local in Winter only.

RESIDENT SPECIES IN	ENGLAND.	WALES.	SCOTLAND.	IRELAND.
Turdus musicus (Song Thrush)	×	×	×	×
Turdus viscivorus (Missel Thrush) ...	×	×	×	×
Merula merula (Blackbird)	×	×	×	×
Pratincola rubicola (Stonechat)	×	×	×	×
Erithacus rubecula (Robin)	×	×	×	×
Regulus cristatus (Goldcrest)	×	×	×	×
Sylvia provincialis (Dartford Warbler)	×			
Accentor modularis (Hedge Accentor)	×	×	×	×
Cinclus aquaticus (Dipper)	×	×	×	×
Panurus biarmicus (Bearded Titmouse)	×			
Acredula caudata rosea (Long-tailed Titmouse)	×	×	×	×
Parus major (Great Titmouse)	×	×	×	×
Parus ater *et* ater britannicus (Coal Titmouse) ...	×	×	×	×
Parus palustris (Marsh Titmouse)	×	×	VL	VL
Parus cæruleus (Blue Titmouse)	×	×	×	×
Parus cristatus (Crested Titmouse)			×	
Sitta cæsia (Nuthatch)	×	×	L	
Troglodytes parvulus (Wren)	×	×	×	×
Certhia familiaris (Creeper)	×	×	×	×
Motacilla yarrellii (Pied Wagtail)	×	×	×	×
Motacilla sulphurea (Gray Wagtail)	×	×	×	VL
Anthus pratensis (Meadow Pipit)	×	×	×	×
Anthus obscurus (Rock Pipit)	×	×	×	×
Coccothraustes vulgaris (Hawfinch)	×			
Fringilla carduelis (Goldfinch)	×	×	×	×
Fringilla chloris (Greenfinch)	×	×	×	×
Fringilla spinus (Siskin)	×	×	×	×
Fringilla cœlebs (Chaffinch)	×	×	×	×
Linota cannabina (Linnet)	×	×	×	×
Linota rufescens (Lesser Redpole)	×	×	×	×
Linota flavirostris (Twite)...	×	×	×	×
Passer domesticus (House Sparrow)	×	×	×	×
Passer montanus (Tree Sparrow)...	×	L	L	L
Pyrrhula vulgaris (Bullfinch)	×	×	×	×
Loxia curvirostra (Crossbill)	×	×	×	×
Emberiza miliaria (Common Bunting)	×	×	×	×
Emberiza citrinella (Yellow Bunting)	×	×	×	×
Emberiza cirlus (Cirl Bunting)	×			
Emberiza schœniclus (Reed Bunting) ...	×	×	×	×
Sturnus vulgaris (Starling)	×	×	×	×
Pyrrhocorax graculus (Chough)	×	×	×	×
Garrulus glandarius (Jay)	×	×	×	VL
Pica caudata (Magpie)	×	×	×	×
Corvus monedula (Jackdaw) ...	×	×	×	×
Corvus corax (Raven)	×	×	×	×
Corvus corone (Carrion Crow) ...	×	×	×	VL
Corvus cornix (Hooded Crow) ...	W	W	×	×

SPECIES RESIDENT IN	ENGLAND.	WALES.	SCOTLAND.	IRELAND.
Corvus frugilegus (Rook)	×	×	×	×
Alauda arvensis (Sky Lark)	×	×	×	×
Alauda arborea (Wood Lark)	×	×	L	L
Gecinus viridis (Green Woodpecker)	×	×		
Picus major (Great Spotted Woodpecker) ...	×	×		
Picus minor (Lesser Spotted Woodpecker) ...	×	×		
Alcedo ispida (Kingfisher)	×	×	×	×
Aluco flammeus (Barn Owl)	×	×	×	×
Asio otus (Long-eared Owl)	×	×	×	×
Asio brachyotus (Short-eared Owl)	×	×	×	W
Strix aluco (Tawny Owl)	×	×	×	
Circus æruginosus (Marsh Harrier)	×	×	×	×
Circus cyaneus (Hen Harrier)	×	×	×	×
Buteo vulgaris (Common Buzzard)	×	×	×	×
Aquila chrysaëtus (Golden Eagle)	W	W	×	×
Haliaëtus albicilla (White-tailed Eagle) ...	W	W	×	×
Accipiter nisus (Sparrow Hawk)	×	×	×	×
Milvus regalis (Kite)	×	×	×	
Falco æsalon (Merlin)	×	×	×	×
Falco tinnunculus (Kestrel)	×	×	×	×
Falco peregrinus (Peregrine)	×	×	×	×
Phalacrocorax carbo (Cormorant)	×	×	×	×
Phalacrocorax graculus (Shag)	×	×	×	×
Sula bassana (Gannet)	×	×	×	×
Ardea cinerea (Heron)	×	×	×	×
Anser cinereus (Gray-Lag Goose)	×		×	×
Tadorna cornuta (Sheldrake)	×	×	×	×
Anas boschas (Mallard)	×	×	×	×
Anas clypeata (Shoveller)	×	×	×	×
Anas crecca (Teal)	×	×	×	×
Somateria mollissima (Eider Duck)	VL		×	
Mergus serrator (Red-breasted Merganser) ...	W	W	×	×
Columba palumbus (Ring Dove)	×	×	×	×
Columba œnas (Stock Dove)	×	×	L	VL
Columba livia (Rock Dove)	×	×	×	×
Tetrao urogallus (Capercaillie)	×	×	×	×
Tetrao tetrix (Black Grouse)	×	× ?	×	
Lagopus mutus (Ptarmigan)			×	
Lagopus scoticus (Red Grouse)	×	×	×	×
Phasianus colchicus (Pheasant)	×	×	×	×
Perdix cinerea (Partridge)	×	×	×	×
Rallus aquaticus (Water Rail)	×	×	×	×
Gallinula chloropus (Waterhen)	×	×	×	×
Fulica atra (Coot)	×	×	×	×
Ægialitis hiaticula major (Greater Ringed Plover)	×	×	×	×
Charadrius pluvialis (Golden Plover)	×	×	×	×
Vanellus cristatus (Lapwing)	×	×	×	×
Hæmatopus ostralegus (Oystercatcher) ...	×	×	×	×
Scolopax rusticola (Woodcock)	×	×	×	×

SPECIES RESIDENT IN	ENGLAND.	WALES.	SCOTLAND.	IRELAND.
Scolopax gallinago (Snipe)	×	×	×	×
Tringa alpina (Dunlin)	×	×	×	×
Totanus calidris (Redshank)	×	×	×	×
Numenius arquata (Curlew)	×	×	×	×
Larus ridibundus (Black-headed Gull)	×	×	×	×
Larus canus (Common Gull)	W	W	×	×
Larus argentatus (Herring Gull)	×	×	×	×
Larus fuscus (Lesser Black-backed Gull) ...	×	×	×	×
Larus marinus (Greater Black-backed Gull) ...	×	×	×	×
Larus tridactylus (Kittiwake)	×	×	×	×
Stercorarius catarrhactes (Great Skua)	W	W	×	VLW
Alca torda (Razorbill)	×	×	×	×
Uria troile (Guillemot)	×	×	×	×
Uria grylle (Black Guillemot)	VL	W	×	×
Podiceps cristatus (Great-crested Grebe) ...	×	×	×	×
Podiceps minor (Little Grebe)	×	×	×	×
Puffinus anglorum (Manx Shearwater)	×	×	×	×
Procellaria leachi (Fork-tailed Petrel)	×	×	×	×
Procellaria pelagica (Stormy Petrel)	×	×	×	×

NOTE.—× signifies Breeding; VL Breeding, but very locally; L ditto, but locally; W Winter only; VLW Very rare and local in Winter only.

Of the 115 species, tabulated above, that are resident in our area, fourteen are entirely absent from Ireland, whilst another eight cannot in any sense be considered dominant in the island, but are excessively local. Now it is a most significant fact that all these fourteen species (with the solitary exception of the Eider Duck) are confined chiefly to England and the extreme south of Scotland, eminently southern species, not ranging north of France in some cases, not north of the Baltic in others; or in the few cases of wide range northwards they are birds strictly sedentary in their habits, and thus leave no indication of former residence in Ireland (if such were ever the case) by a line of Migration thereto. Further, there can be no question whatever that all these species were late arrivals, comparatively speaking, in the British

Area, that their emigrations northwards and westwards did not reach the west of England before the isolation of Ireland had taken place, and a barrier to extension in that direction formed; a fact which is absolutely proved by the line of Emigration followed by those species, which it will be found in no single case reached Scandinavia by way of Britain. On the other hand, nearly half of the species (45 out of 99) resident in Ireland at the present time emigrated to Scandinavia by way of the British Islands. The most probable explanation of the one solitary exception of the Eider Duck is the fact that Ireland is situated too far south, and that suitable groups of small islets are not present off the coast. Most of the resident species absent from Ireland are birds that become rarer and more local in the west of England or in Wales, and in at least eight cases they are also absent from or very local in Scotland. But three species are resident in Scotland and not in England, all thoroughly hardy forms (*Parus cristatus*, *Corvus cornix*, and *Lagopus mutus*); five species breed in Scotland and Ireland, but not in England (although at least three formerly did so). Seven species breed only in England and Wales, but no species breeds exclusively in Ireland. Of the 115 resident species in the British Isles, 108 breed in Scotland, 107 in England, or in England and Wales, and 99 in Ireland.

Still more significant are the facts presented by the Summer Migrants to the British Islands. These are tabulated on the following page.

NOTE.—× signifies Breeding; L Locally; VL Very Locally; M On Migration; VLM Very Locally on Migration; LM Locally on Migration; M× Chiefly on Migration, few Breeding.

SUMMER MIGRANTS TO	ENGLAND.	WALES.	SCOTLAND.	IRELAND.
Merula torquata (Ring Ouzel)	×	×	×	×
Saxicola œnanthe (Wheatear)	×	×	×	×
Pratincola rubetra (Whinchat)	×	×	×	×
Ruticilla phœnicurus (Redstart)	×	×	×	
Erithacus luscinia (Nightingale)	×	×		
Sylvia cinerea (Whitethroat)	×	×	×	×
Sylvia curruca (Lesser Whitethroat)	×	×	×	
Sylvia atricapilla (Blackcap)	×	×	L	VL
Sylvia hortensis (Garden Warbler)	×	×	×	VL
Phylloscopus rufus (Chiffchaff)	×	×	×	L
Phylloscopus trochilus (Willow Wren)	×	×	×	×
Phylloscopus sibilatrix (Wood Wren)	×	×	×	VL
Acrocephalus arundinaceus (Reed Warbler) ...	×	×		
Acrocephalus palustris (Marsh Warbler) ...	×			
Acrocephalus phragmitis (Sedge Warbler) ...	×	×	×	×
Locustella locustella (Grasshopper Warbler) ...	×	×	×	L
Locustella luscinioides (Savi's Warbler)... ...	×			
Motacilla raii (Yellow Wagtail)	×	×	L	VL
Motacilla alba (White Wagtail)	×	×?	×	
Anthus trivialis (Tree Pipit)	×	L	×	
Oriolus galbula (Golden Oriole)	×			
Lanius collurio (Red-backed Shrike) ...	×	×?	VL	
Muscicapa grisola (Spotted Flycatcher) ...	×	×	×	×
Muscicapa atricapilla (Pied Flycatcher) ...	×	×	L	
Hirundo rustica (Swallow)	×	×	×	×
Chelidon urbica (Martin)	×	×	×	×
Cotyle riparia (Sand Martin)	×	×	×	×
Cypselus apus (Swift)	×	×	×	×
Caprimulgus europæus (Nightjar) ...	×	×	×	×
Iynx torquilla (Wryneck)	×	L		
Upupa epops (Hoopoe)	×			×
Cuculus canorus (Cuckoo)...	×	×	×	×
Circus cineraceus (Montagu's Harrier) ...	×	×		
Falco subbuteo (Hobby)	×			
Pernis apivorus (Honey Buzzard)... ...	×			
Pandion haliaëtus (Osprey)	M	M	×	M
Ardetta minuta (Little Bittern)	×			
Platalea leucorodia (Spoonbill) *extinct* ...	×			
Anas circia (Garganey)	×			×
Turtur auritus (Turtle Dove)	×	×	×	×
Coturnix communis (Quail)	×	×	×	×
Crex pratensis (Land Rail)	×	×	×	×
Crex porzana (Spotted Crake)	×	×	×	×
Grus communis (Crane) *extinct*	×			VLM
Œdicnemus crepitans (Stone Curlew) ...	×			
Eudromias morinellus (Dotterel)... ...	M	M	×	VLM

SUMMER MIGRANTS TO	ENGLAND.	WALES.	SCOTLAND.	IRELAND.
Ægialophilus cantianus (Kentish Plover) ...	×			
Recurvirostra avocetta (Avocet)	×			
Phalaropus hyperboreus (Red-necked Phalarope)	LM			
Totanus pugnax (Ruff)	×			× LM
Totanus hypoleucus (Common Sandpiper) ...	×		×	×
Totanus glareola (Wood Sandpiper)	×		× ?	
Totanus ochropus (Green Sandpiper)	× ?	× ?	M	× LM
Limosa melanura (Black-tailed Godwit)... ...				
Numenius phæopus (Whimbrel)	M	M	M ×	M
Hydrochelidon nigra (Black Tern) *extinct* ...	×			
Sterna cantiaca (Sandwich Tern)...	×		×	
Sterna dougalli (Roseate Tern)	×		×	
Sterna hirundo (Common Tern)	×	×	×	
Sterna arctica (Arctic Tern)	×	×		
Sterna minuta (Lesser Tern)	×			×
Stercorarius richardsoni (Richardson's Skua)	M	M		M
Fratercula arctica (Puffin)...	×	×		×

As will be seen from the table above given these number 63 species, but two of these, however (the Osprey and the Dotterel), have now ceased to breed in England, although they did so say within the last hundred years, and three others (the Red-necked Phalarope, the Whimbrel and Richardson's Skua) now only pass that country on migration, but there is no doubt that in earlier ages, when the climate was sufficiently Arctic, they bred there. Four of these species also pass Ireland on migration, which is a safe indication that they formerly occupied that area as breeding species when extending their range northwards after the Glacial Epoch had passed. Exclusive of these species, but including the three that have become extinct purely through man's interference, we find that out of the 58 Summer Migrants to the British Islands no less than 26, or nearly one-half, have either never been detected in Ireland at all, or have only occurred there as purely abnormal wanderers! Six others migrate to Ireland every spring, but are very locally distributed in the island, and in most cases confined to the eastern and southern districts. It is also interesting to remark that all these absent species were late arrivals in our area (presumably when Ireland had become isolated from England), a fact confirmed by the absence of any trace of their emigration northwards to Scandinavia across England and Scotland; whilst three out of the four species that do visit Ireland very locally on passage undoubtedly extended their range *viâ* our area to Scandinavia, viz. the Crane, the Dotterel, and the Ruff (*conf.* table, p. 122). It is also a significant fact that not one of the three Summer Migrants that visit us from the south-east (*conf.* table, p. 84) has been known to visit Ireland, except the Red-backed Shrike, a solitary example of which was obtained near Belfast on abnormal flight. All but five of these Summer Migrants to the British Islands breed in England, and of these five, two did so within comparatively recent years, viz. the Osprey and the Dotterel. Twenty species are also absent from Scotland, but these are all southern forms whose dominant range in England itself is not very northerly. All the species that visit Ireland also range further north into Scotland, with the exception of the Garganey and the Hoopoe, which are both decidedly southern types; and with these two exceptions all the Summer Migrants to Ireland are dominant wide-ranging species over the whole of the British Area. It is to be deplored that we know so little of the distribution of birds in Ireland. Correct detailed information is much to be desired, and will materially assist us in our study of the Emigration and Migration of birds within the British Archipelago.

We will next deal with the species that either visit the British Islands in autumn and pass the winter therein, or migrate across them in spring and autumn to winter quarters further south, and breeding-grounds further north. These species are tabulated below.

NOTE.—A few individuals of some of the species (marked *) breed in our area, so that some of them may be said to be *resident* therein, but for obvious reasons they have not all been included with our typical residents. W signifies Winter Visitor; RW Rare Winter ditto; LW Local Winter ditto; MW Migration and Winter, chiefly the former; ×M Migration chiefly, few Breed; ×W Winter chiefly, few Breed; CM Coasting Migrant; RM Rare on Migration; AM Abnormal Migrant.

AUTUMN MIGRANTS TO, OR COASTING MIGRANTS OVER	ENGLAND.	WALES.	SCOTLAND.	IRELAND.
Turdus iliacus (Redwing)	W	W	W	W
Turdus pilaris (Fieldfare)	W	W	W	W
Ruticilla tithys (Black Redstart) ...	W	W	RW	RW
Regulus ignicapillus (Firecrest) ...	W	LW	LW	
Lanius major (Great Gray Shrike) ...	W	W	W	LW
Ampelis garrulus (Waxwing)	W	W	W	RW
* Plectrophenax nivalis (Snow Bunting)... ...	W	W	×W	W
Fringilla montifringilla (Brambling)	W	W	W	W
Otocoris alpestris (Shore Lark)	W	LW	LW	
Nyctea nyctea (Snowy Owl)	W		W	W
Archibuteo lagopus (Rough-legged Buzzard) ...	W	W	W	AM
Astur palumbarius (Goshawk)	W	W	W	AM
Hierofalco candicans (White Jer Falcon) ...	W	W	W	W
Hierofalco islandus (Iceland Jer Falcon) ...	W	?	W	W
Hierofalco gyrfalco (Scandinavian Jer Falcon) ...	W			
Anser albifrons (White-fronted Goose)	W	W	W	W
Anser segetum (Bean Goose)	W	W	W	W
Anser brachyrhynchus (Pink-footed Goose) ...	W	W	W	
Bernicla leucopsis (Bernacle Goose)	W	W	W	W
Bernicla brenta (Brent Goose)	W	W	W	W
Cygnus musicus (Hooper Swan)	W	W	W	W
Cygnus bewicki (Bewick's Swan)...	W	W	W	W
Cygnus olor (Mute Swan)...	W	W	W	
* Anas strepera (Gadwall)...	×W	W	W	W
* Anas acuta (Pintail Duck)	W	W	×W	×W
* Anas penelope (Wigeon)	W	W	×W	×W
* Fuligula ferina (Pochard)	×W	W	×W	×W

AUTUMN MIGRANTS TO, OR COASTING MIGRANTS OVER	ENGLAND.	WALES.	SCOTLAND.	IRELAND.
* Fuligula cristata (Tufted Duck)	× W	W	× W	× W
Fuligula marila (Scaup)	W	W	W	W
Fuligula glacialis (Long-tailed Duck)	W	W	W	W
* Fuligula nigra (Common Scoter)	× ?W	W	× W	W
Fuligula fusca (Velvet Scoter)	W	W	W	W
Clangula glaucion (Golden Eye)	W	W	W	W
* Mergus merganser (Goosander)...	W	W	× W	LW
Mergus albellus (Smew)	W	W	W	RW
Botaurus stellaris (Bittern)	W	W	W	RW
Otis tarda (Great Bustard)	W	W	W	RW
Otis tetrax (Little Bustard)	W		W	RW
Ægialitis hiaticula (Ringed Plover)	CM	CM	CM	CM
Charadrius helveticus (Gray Plover)	MW	MW	MW	MW
Strepsilas interpres (Turnstone)	W	W	W	W
Phalaropus fulicarius (Gray Phalarope)	W	RW	RW	RW
† Scolopax major (Great Snipe)	CM		RM	RM
Scolopax gallinula (Jack Snipe)	W	W	W	W
† Tringa minuta (Little Stint)	CM	CM	CM	CM
† Tringa temmincki (Temminck's Stint) ...	RM	RM	RM	
† Tringa subarquata (Curlew Sandpiper) ...	CM	CM	CM	CM
Tringa maritima (Purple Sandpiper)	W	W	W	W
Tringa canutus (Knot)	MW	MW	MW	MW
Tringa arenaria (Sanderling)	CM	CM	CM	CM
† Totanus fuscus (Dusky Redshank)	RM		RM	RM
* Totanus glottis (Greenshank)	CM	CM	× M	MW
Limosa rufa (Bar-tailed Godwit)...	W	W	W	W
Larus minutus (Little Gull)	RW	RW	RW	RW
Larus glaucus (Glaucus Gull)	W	W	W	RW
Pagophila eburnea (Iceland Gull)	RW	RW	RW	RW
Stercorarius buffoni (Buffon's Skua)	CM		CM	CM
Stercorarius pomatorhinus (Pomatorhine Skua)	W	W	W	RW
Mergulus alle (Little Auk)	W	W	W	W
Colymbus glacialis (Great Northern Diver) ...	W	W	W	W
* Colymbus arcticus (Black-throated Diver) ...	W	W	× W	W
* Colymbus septentrionalis (Red-throated Diver)	W	W	× W	× W
Podiceps rubricollis (Red-necked Grebe) ...	W	W ?	W	
Podiceps cornutus (Sclavonian Grebe)... ...	W	W	W	W

† *Conf.* pp. 87, 88, 91, 92.

When we come to study the species that visit the British Islands in winter, or pass them on migration, a very different state of things occurs. The birds that come into this class number 64 species. Of these the Firecrest, the Shore Lark, the Scandinavian Jer Falcon, the Mute Swan, the Pink-footed Goose, Temminck's Stint, and the Red-necked Grebe are as yet unknown in Ireland; but as at least four of these birds are extremely uncommon in all other parts of the British Area, and the remaining three have a strongly-marked eastern tendency in their migrations, it is scarcely fair to take them into consideration in our analysis. Of the 57 species that visit Ireland in winter, or pass over it on migration to and from the north, it is a very suggestive fact that no less than 29 breed in Greenland, Iceland, or the Faroes, or pass the Faroes and Iceland on migration—the nearest land masses north of the Irish Area lying beyond the British limits. Of the remaining 28 species, at least four (Bewick's Swan, Bean Goose, Pomatorhine Skua, and Buffon's Skua) spend the summer in the high north in little known lands between Greenland and Nova Zembla; whilst twelve of the rest breed more or less commonly in Scandinavia, and whose migrations to our area may be traced in all but one instance (*Colymbus arcticus*), by way of the Shetlands and Orkneys. Four others are known or presumed to breed in Arctic Russia (the Smew, the Gray Plover, the Curlew Sandpiper, and the Bar-tailed Godwit), but in the case of the three latter more western breeding grounds may yet be discovered; whilst the Smew is decidedly rare in Ireland. Of the remaining eight species two (the Rough-legged Buzzard and the Goshawk) are only abnormal wanderers to Ireland, and six are decidedly rare in any part of the United Kingdom, viz. the Black Redstart, the Waxwing, the Bittern, the Great Bustard, the Little Bustard, and the Little Gull (all rarer in Ireland even than elsewhere), species whose line of Migration in autumn is more southerly than westerly, and whose breeding grounds in some cases are for the most part below the parallels of Irish latitude, and in others whose centre of dispersal was from the south-east.

These facts prove undoubtedly that the common and dominant species that exclusively winter in Ireland and in other portions of the British Islands, or that pass such areas on migration, are strictly boreal forms, most of them from the high north or north-east; the few that reach us from the east are mostly rare and irregular, and confine their visits principally to the eastern side of our islands. Of course I need scarcely state that I do not include our normal east to west migrants (*conf.* table, p. 132) in these remarks. The proportions of the Irish avifauna to that of England, Wales, and Scotland may be thus briefly summarized.

CLASSIFICATION.	BREEDING IN ENGLAND AND SCOTLAND.	BREEDING IN IRELAND.
Summer migrants, 63.	63.	32.
Resident species, 115.	115.	99.
Winter visitors or coasting migrants: 64, less 6 better classed as abnormal migrants; total 58.	Wintering in or passing over England and Scotland: 58.	Wintering in or passing over Ireland: 57, less 2 abnormals; total 55.

The significance of this proportion is apparent when we study it in relation to the following facts. It may be stated as an axiom, that as species begin to extend their breeding or summer range northwards they still continue to return to winter in the area which they formerly occupied as residents, until such time as that extended summer area may be dwelt in permanently owing to an amelioration of climate ; and that as the range becomes more and more northerly, and less and less southerly, the intermediate areas are only passed on migration or occupied by comparatively small numbers of individuals in summer and winter—a few individuals breeding in the area, a few wintering ; but even then there is every probability that the former draw south in winter, and are replaced by individuals breeding further north. We may thus assume that the species that pass our islands on migration were the earliest to enter our area after the Glacial Epoch, and have followed the ameliorating climate northwards, none of them breeding in our islands now. We may also assume that the species that return to our islands to winter therein are the descendants of birds that were somewhat later in emigrating northwards after glacial conditions had lapsed, and as is usual in such cases a few individuals remain to breed in our area—direct descendants, we must regard them, of the very last individuals that moved north to that area, and which we can have no doubt in many if not in all cases move south of our islands to winter.

That both the resident species in the British Area, as well as those that winter therein, or pass over it on migration, were early emigrants is still further confirmed by the fact of their being so dominantly distributed in Ireland. They reached that area when it was joined to England in the south, and before much of the land was submerged that is now covered by St. Georges Channel, the Bristol Channel, and the English Channel. They continue to winter in Ireland in such proportionate abundance probably because the land continues to be much more compact and continuous in the north (the point of egress) than it is now in the south. On the other hand, there is much evidence to suggest that the stream of Coasting Migrants across the Irish Area is much less in volume than the stream that passes over Scotland and England, which must be due in a great measure to the wide water areas south of Ireland. Many instances, moreover, might be given in which a number of individuals of certain species, which pass Scotland and England only on migration, spend the winter in the south of Ireland, as though their southern progress was arrested by the impassable barrier of the open Atlantic Ocean. Thus many Knots, Greenshanks, Gray Plovers, Bar-tailed Godwits, and Short-eared Owls pass the winter in Ireland; whilst even such typical summer migrants as the Blackcap Warbler and the Land Rail, for instance, are frequently known to do so in the south of that island—Coasting Migrants from the north which have apparently gone too far west, and whose southern flight has been abnormally arrested by the wide area of unknown sea. We often hear of similar phenomena in the south-west of England, where the sea-passage to the Continent is very wide. I am not disposed to consider such instances as due to the milder climate of these areas tempting a winter sojourn: the individual birds in question are lost migrants, too far west of their normal route.

The Summer Migrants represent the species that were the last to extend their breeding range northwards across the British Area—then a portion of continental land. As many of these birds reached the British Islands, we have every reason to believe that submergence had commenced, and that Ireland had become separated from England long before it had received its fair share of the birds that were spreading across our land as summer visitors. Thus we have birds that visit England, and even extend their summer flight to the north of Scotland, such as the Redstart, the Lesser Whitethroat, and the Tree Pipit, but which do not visit Ireland, which they normally should do, all conditions being equal; the water areas forming an impassable barrier to extension of range thereto. These species also furnish another proof in support of the Law which forbids emigration southwards, as in the north-east of Ireland the water passage to Scotland is no wider than the Strait of Dover between England and France, which these summer birds must cross and re-cross each year, and therefore no barrier to emigration to Ireland, if such a line of extension were a normal one. Ireland is often held up as an example of zoological poverty, and rightly so, in comparison with adjoining areas, but so far as birds are concerned this poverty only exists among one class of species alone, the Summer Migrants to our area, which were prevented from entering the country in past ages owing to submergence of the connecting land areas, and because such species were late emigrants in this direction, arriving in the British Area, we have every reason to believe, long after the dominant resident avifauna was established, or the present winter visitors thereto and coasting migrants across had spread northwards over it.

In order to render the subject of Avian Emigration to the British Area fairly complete, I append below a table of the species and races that are peculiar to this area, and which I propose to discuss in greater detail in a later chapter (*conf.* p. 192 *et seq.*).

BRITISH RACE.	BRITISH DISTRIBUTION.	CONTINENTAL OR EXTRA-BRITISH PARENT SPECIES.	POINTS OF DISTINCTION OF BRITISH RACE.
Parus ater britannicus	General.	Parus ater.	Upper parts below nape, and the flanks, more suffused with brown.
Parus palustris dresseri	General in England and Wales; local in Scotland and Ireland.	Parus palustris.	General colouration more dull, and more suffused with sandy brown.
Acredula caudata rosea	General.	Acredula rosea.	General colouration more dull; white on head confined to crown; lores black.
Troglodytes parvulus hirtensis	St. Kilda.	Troglodytes parvulus.	Upper parts below nape barred with dark brown; bill and feet larger.
Lagopus scoticus	General.	Lagopus albus.	Quill feathers brown at all seasons; no white winter plumage.

A brief *résumé* of the facts (and the deductions drawn from them) contained in the preceding and the present chapter will perhaps help to simplify what is undoubtedly a very intricate subject, and enable the reader better to grasp the importance of their bearing upon the question of avian dispersal. We commenced Chapter III. by propounding a Law of Dispersal, which I think very closely governs the facts set forth in that and the present chapter. By means of this Law and its corollaries we are able to explain many apparent anomalies of avian distribution. For instance, they will explain why the Arctic Tern actually breeds in Alaska, yet returns to South-eastern Asia to winter, not breeding on any other part of the Pacific shores of the American continent, because such areas could only be reached by a flight south in spring. It explains why the Little Bunting breeds in Siberia within the same parallels of latitude as the British Islands, and only in North Russia in Europe; why the Eastern Nightingale breeds from Asia Minor to South Sweden, yet is unknown in Western Europe; why the Red-backed Shrike breeds throughout the whole of Europe south of lat. 64°, with the sole exception of the Iberian Peninsula; why the Nightingale, the Reed Warbler, and other birds are absent from the south-west of England; why *Turdus aliciæ, Junco hyemalis, Ægialitis semipalmatus*, and *Tryngites rufescens* cross over into North-eastern Siberia to breed yet return to their range bases in America to winter; why *Phylloscopus borealis* and *Erithacus suecica*, after spending the summer in Alaska, return to winter exclusively in the Old World; together with scores of other anomalies of distribution which space alone forbids me to enumerate. I appeal to naturalists interested in that fascinating branch of ornithology which embraces Geographical Distribution and Migration to apply this Law; for I am confident that it will lead to important results, and explain many anomalies; perhaps even enable us to remodel on a more satisfactory and natural basis the various regions into which the world has been divided by zoologists.

We began with the vast change of climate that marked the closing era of the Pleistocene Period, and endeavoured to trace its effects upon the avifauna in particular of Western Europe; we have sketched out the probable condition of the British Area during this cold era, and attempted an outline of the probable resident avifauna of that and adjoining areas during this period. We traced out the two other Range Bases or Refuge Areas during the close of the Glacial Epoch of most of the species that now compose the avifauna of West Europe, by the aid of their allied southern and parent races or representative species left stranded in Iberia, North-west Africa, the Canary Islands, Malta, etc. We have also sought to demonstrate not only that the range of a very large percentage of our resident species is continuous from our area down to Iberia, North-west Africa, and the Canary Islands; but also that a by no means inconsiderable number of species resident in our islands or met with in them throughout the year resort to that region (Refuge Area II.) in winter, as very conclusive evidence of the source from which our avifauna was derived. Similar confirmation of this fact has also been shown to exist among the species that are only known as winter visitors to our area, many of them extending their range at that season to varying limits as far south as the Canaries. We have also shown that almost all our Summer Visitors seek winter quarters in this Refuge Area (II.). Some of these species go much further south in the east than in the west; some of them breed as well as winter in North-west Africa (a most convincing proof that that was an area of dispersal or base from which their emigrations northwards commenced), yet only pass North-east Africa on migration—two apparent anomalies for which I think a satisfactory reason has been given. We have also dealt with the two small groups of

Summer Migrants to our area, the species in one of which apparently reach us by way of Algeria and Iberia (or perhaps Italy and France); the species in the other group by way of a north-west migration across Europe, as is proved by their absence from North-west Africa and Iberia. We have then proceeded to deal with another important class of birds—nomadic migrants and abnormal wanderers to the British Area—and to ascertain the causes which have prevented these species reaching our islands during past ages, and becoming a portion of its dominant avifauna, and the relation of these facts to past geographical conditions. We then approached the difficult problem presented by birds inhabiting West Europe—even in some cases within sight of our shores—yet absent from the British Area, and by an application of our Law of Dispersal to many instances of such avian distribution, have shown that the curious facts are perfectly normal.

Coming then to the subject of wide water areas as most effectual barriers to Emigration, and its bearing on the dispersal of British birds, we have considered at some length the vast and far-reaching influence of such a barrier on the avifauna of our area, and have endeavoured to show why so many species breeding in West Europe never attempt to extend their range across the English Channel. Coming to the question of a former land connection between Greenland and Europe, we have dealt with the causes of the ornithological poverty of Greenland, the reason why the avifauna of Iceland does not contain more Nearctic elements, and have endeavoured to demonstrate the importance of the emigration that in long past ages followed this route to the Arctic regions.

In the beginning of the present chapter we have discussed the possibility of undiscovered breeding grounds between Greenland and Spitzbergen of several northern birds, as an explanation of various apparent anomalies of distribution; and by an analysis of the table containing the species that emigrated in a north-westerly direction to Iceland and Greenland, we have shown what a great and significant proportion of species continue to breed along the entire route from the British Area to Greenland, and a still greater proportion between Iceland and our area. We have also shown that in two cases at least the highest northern emigration of the species throughout the world has been along this route. We next proceed to discuss the two dominant lines of Post-Glacial Emigration in West Europe, illustrating by a table the range extension northwards of 206 species, and their relation to the British avifauna. By means of this inquiry we discover the astonishing fact that a vast percentage of species range absolutely higher on the Continent than they do in our area, notwithstanding its much milder climatal conditions—a discrepancy of distribution due entirely to the dominant line of Emigration followed in past ages, and a fact which proves that the individuals of certain species breeding in North-west Europe (say in Scandinavia) are descendants of the individuals that dwelt in the south-east (Refuge Area III.); whilst individuals of the same species breeding in our islands, and as far north as Holland or Denmark, are the descendants of other individuals that dwelt in the south (Refuge Area

II.); that in fact the avifauna of these two portions of West Europe have been derived from two quite distinct and widely-separated areas. We have then proceeded to deal with the emigration of species eastwards and north-eastwards into continental Europe that started from the British Area as a basis, and which is shown to be recapitulated even at the present time in the migrations of various birds to and from our area across the North Sea. We thus see that Emigration is only attempted across land areas or very narrow water passages, and that in every case where Migration at the present time crosses wide stretches of sea it is an unfailing sign of comparatively recent land submergence, and a sure indication that when the migration across such areas was being established, the land masses were continuous, or nearly so.

Passing on to a more detailed study of the distribution of birds within the British Archipelago, we have found the same Laws of Dispersal to apply, and the facts to be thoroughly in harmony with those climatal and geographical changes which have taken place in this area during Post-Glacial time. We have discussed at some length the impossibilities of southern dispersal, and shown how the views held by Professors Geikie and Leith Adams are opposed to known facts, and that the evidence furnished by birds, animals, insects, and plants leads to the inevitable conclusion that normal extension in the Northern Hemisphere is never pursued in a southerly direction.

We have then proceeded by the aid of a series of tables to demonstrate the emigration of birds to the various portions of the British Area, dealing with the resident species in, the summer and winter migrants to, and the coasting migrants across that area, and to show the proportionate avifauna of each of the three divisions of the United Kingdom. Dealing then with the resident species and each class of Migrant, we have endeavoured to trace the sequence of their arrival in the British Area; and, finally, to append a table of the species and races of birds that are peculiar to that area. The whole subject is a most fascinating one, and full of promise; an unworked field to which I am afraid the somewhat small limits of the present work have prevented me from devoting a fuller measure of attention. Sufficient, I hope, has been discussed to serve as a stimulus to further inquiry. The new Law of Dispersal, by which I have attempted to explain what up to the present time have been universally regarded by naturalists as inexplicable phenomena, will, I trust, meet with some small portion of tentative approval from biologists.

As we remarked in our *résumé* at the close of Chapter II., so we may again repeat here, that the Glacial Epoch was the cause of extermination of the greater portion of the fauna and flora of northern and temperate Euro-Asia. Among this awful wreck of life we must undoubtedly include the Birds. Some of these species, as we know from palæontological evidence, must have disappeared from the scene for ever; others, the majority of others, have left no trace behind them. Birds, it may be urged, could readily escape from such adverse conditions: as the severe winters came on they could retire south. But all the available evidence we can bring to bear upon this question disproves the assumption. In the Northern Hemisphere no bird increases its area of distribution southwards, either in summer or winter; if it be overtaken by a change to a colder climate in the north, the individuals of the species that occupy that area will assuredly perish, and that portion of the species be eradicated.[1] If, for instance, the British Islands are visited by an exceptionally severe winter, birds, instead of flying off south to escape its terrors, die helplessly in thousands and tens of thousands; and it has been remarked that certain localities have been well-nigh depopulated of various birds, not perhaps regaining their usual strength again for years. If the birds, instead of perishing from cold and hunger, had merely emigrated south a little way (perhaps but a very few miles!) they would have been safe, and the return of spring weather would have brought them back in their wonted numbers again. We have at least some direct evidence that the cold winters we have been so persistently having of late years are slowly exterminating the Dartford Warbler. As a British species it is undoubtedly much rarer than it was twenty years ago. It is sedentary, and if these conditions continue nothing can save this little bird from total extinction in the British Islands. We observe no extension of range southwards, no "re-

[1] A species may become gradually extinct through the inability to rear offspring owing to adverse climatic conditions. In some seasons, for instance, the Brent Goose appears on our coasts unaccompanied by any young; and it may fairly be presumed that an unfavourable season (say abnormally cold, backward, or stormy) in the high latitudes where this species breeds has proved fatal to the eggs or young. Should these adverse climatic conditions become more pronounced, and continue season after season (as they must have done during the coming on of a Glacial Period) the species would most assuredly gradually be exterminated.

treat" to escape the growing inclemency of our English winters. Now if we take the instance of the Dartford Warbler and apply it to a Glacial Epoch, what are the results? The northern individuals would perish; the southern representatives of the species would survive, and by Post-Glacial Emigration would expand their previously contracted area, and re-stock the northern lands wherever conditions were favourable to such extension. Had the species no southern base beyond the limits of glaciation or the effects of the severe climate, *it must have been exterminated.* Did it manage to survive in any favourable locality in the south, that locality must have been within its Pre-Glacial range. With Inter-polar (or perhaps Inter-hemisphere) species, however, the conditions were different. These species were probably always migratory; glaciation might curtail their summer or winter range, and in combination with equinoctial precession entirely reverse it, but their migratory habits would preserve them from extinction. The Glacial Epoch, then, exerted its baneful influence upon all living things that dwelt permanently within its limits, and mercilessly if slowly exterminated all species, or the northern portions of all species, wherever they came within its power. *Such species as had no southern Pre-Glacial breeding base beyond the limits of glacial influence to sustain them vanished for ever; only those well rooted in more southerly latitudes preserved themselves, and from these have descended all the emigrants that have re-peopled the once devastated northern areas.* I may here remark, that by "contraction of range" I mean the complete extermination of all individuals of a species within the area affected,[1] save, of course, in such instances which purely apply to Inter-polar or Inter-hemisphere species. Of course it does not come within our present province to seek to show how Evolution progressed, and the segregation of species went on throughout this awful era, but that it was a means of much differentiation is unquestionable.

[1] I may here take the opportunity of pointing out that of the three small families of birds that have solely survived the Ice Age in the Nearctic and Palæarctic regions (the Alcidæ, Colymbidæ, and Panuridæ), by far the most extensive group—the Alcidæ or Auks—numbering some thirty species survives in greatest abundance *in the region where glacial conditions were the least pronounced!* This family of birds may be said to have its head-quarters now round the coasts of the North Pacific (from California to Japan), where not only the bulk of the species are found, but where it preserves the greatest diversity of its surviving development. There can be little doubt that before the Glacial Epoch the Alcidæ were a large circumpolar family; during that era they suffered almost complete extermination, the only survivors being those species or their ancestral forms whose breeding range extended beyond the adverse influences associated with that period. By far their most important range base during the Ice Age must have been the Asiatic coasts and islands of the North Pacific, and possibly the Aleutian Islands—areas, be it remarked, that remained for the most part free from glaciation. A few species, as we have already noted, survived in the North Atlantic, owing partly, as we have abundant evidence to suggest, to the fact of their pre-glacial breeding range, or a portion thereof in the east, having been in the area I have designated Range Base I. (*conf.* pp. 65-77).

CHAPTER V.

RECENT EMIGRATION.

Increase, the Ruling Impulse of Life—The Ice Age and Emigration—Emigration still in Progress—Effects of Civilization on the Emigration of Birds—Present Emigration in the British Area—Emigration of the Missel Thrush—Effects of Severe Winter on that Bird—Emigration of Song Thrush and Blackbird—Of the Redstart—Of the Robin, the Nightingale, the Whitethroat, the Willow Wren, and the Wood Wren—Absence of Wood Wren from Norway—Probable Winter Quarters of Individuals Breeding in Sweden—Emigration of Marsh and Sedge Warblers—Of the Goldcrest—Migration Waves of Goldcrests—Emigration of Hedge Accentor, Nuthatch, and Tree Pipit—Of the Greenfinch—Fluctuating Breeding Range of this Species in the British Area—Emigration of Sparrows—Emigration of Tree Sparrow to the Faroes—Emigration of Chaffinch and Bullfinch—Of the Starling, the Jay, the Magpie, and the Rook—Of the Tawny Owl, the Ring Dove, and the Stock Dove—Of the Great Crested Grebe and Woodcock—Table of Species whose Emigrations are still in Progress—Analysis of Table—Northward Tendency of Emigration—Emigration attended by Migration—Extinction of British Species—Effects of Law of Dispersal.

THROUGHOUT the preceding chapters we have endeavoured to show upon what a vast scale the emigration of birds took place, from the close of the Glacial Epoch down to comparatively recent time. We have seen that birds have spread in endless directions from certain centres, emigrated north, east, and west, in whichever direction an outlet for their increasing numbers was presented, or in which an amelioration of climate, or a return to more suitable conditions permitted an extension of area. The one dominant ruling passion of Life in its endless varying forms is to increase and to spread. That such an impulse to spread still rules supreme, though less acutely than in the more remote Post-Glacial ages, less palpable because the conditions are so much less strongly marked, less dominant than during those chaotic eras—is a fact as incontrovertible as the very existence of Life itself. That the world is never in a state of absolute rest is another truism admitted by every scientific observer of nature. These two facts imply that the Emigration of Life is still in progress; that birds, animals, insects, and plants still continue to colonize new areas, or attempt to do so, with their surplus individuals or excessive and increasing population. This colonizing movement is of a much more restricted nature than it was immediately after the Glacial Epoch, due to the greater difficulties in the way of successful extension of range, owing to the much greater abundance of competing forms, and the much less extensive areas supplying the necessary conditions favourable for such emigration. We can readily understand the impetus given to Emigration by the passing away of the Ice Age, and the opening out of half a hemisphere to the remnants of the species dwelling in the southern areas, vast numbers of which were perhaps dwelling therein at a disadvantage, and under the least favourable conditions for maintaining themselves and for successful increase. We can also understand after the great exodus north was nearly spent, how the stimulus to Emigration would become much less acute, and would therefore progress more slowly, until we can picture the remote future when the northern movement may die completely away either by a complete change to a warm and equable climate at the North Pole, or by the coming on of a new era of glacial conditions. We should, therefore, if the above remarks are correct, expect to observe instances of northern emigration taking place at the present time. This is precisely what we do find. In many parts of the world, especially in countries where owing to a high state of civilization the phenomenon can be carefully observed, Emigration is undoubtedly still going on. If this Emigration is in progress in civilized countries we have every reason to believe that it continues to progress in other parts of the world where the movement cannot be so easily detected; but it may be helped or retarded by the presence or absence of civilization. Thus we have many instances where the spread of cultivation has on the one hand assisted Emigration, whilst on the other hand similar cultivation has had a directly opposite effect, and actually decreased or curtailed the range of many species in certain areas, and, as we know, ultimately banished them from those areas altogether.

In the present chapter I propose to deal exclusively with those species of which we have absolute proof of their emigration being actually in progress, within the limits of the British Archipelago. It will perhaps be best to deal with each species separately, and to give in detail the peculiarities of each emigratory movement, so far as we can trace it, or it has hitherto been observed. The first species whose emigratory movements we will endeavour to trace is the Missel Thrush (*Turdus viscivorus*). This species has very considerably increased its range in our area during the past hundred years. If the early records of the distribution of the Missel Thrush in the British Islands are reliable, we learn that this bird towards the close of the last century was chiefly confined to the lowlands and well-cultivated districts. Whether it was entirely absent from Ireland we have no means of accurately ascertaining, but the probability is that the bird was thinly dispersed over that area as it was in England. Emigration during the past hundred years has been steadily in progress, and at the present time this Thrush is generally distributed over our mainland area. It has also extended its range to the Hebrides, but not yet to the Orkneys. When in Skye some years ago I was informed that the Missel Thrush had become fairly numerous in the island until the severe winter of 1879-80, which almost exterminated it. It is now, however, so I am informed, slowly increasing again in that area. The Song Thrush (*Turdus musicus*) is also extending its range as it increases in numbers in the northern portions of Scotland, and that this fact is not due to the spread of cultivation and tree planting seems proved by the fact of its present rarity in wooded areas of long standing suitable to its requirements. The Blackbird (*Merula merula*) is, comparatively speaking, rapidly extending its range northwards and westwards, and even in some districts encroaching upon the area occupied by the Ring Ouzel, and compelling that species to retire, as in Ross-shire. Macgillivray records that in his day the Blackbird did not breed on the Hebrides; it is now known to nest on them as far north as Lewis. It has also reached the Orkneys, but only visits the Shetlands as an abnormal migrant. The

Redstart (*Ruticilla phœnicurus*) is known to have spread northwards in Scotland of late years, and now breeds sparingly to the very northern limits of the mainland. To Ireland, however, as we have already seen, it is purely an abnormal migrant, which shows how strong a barrier a wide water area is to a successful extension of range. The same remarks apply to the Shetlands; we have many species whose range extends to the Orkneys, but there is little prospect of that range ever normally reaching the more remote group of islets.

In the same way the Robin (*Erithacus rubecula*) has steadily extended its northern range in Scotland, and now breeds regularly in the Hebrides and the Orkneys, where formerly it was excessively rare or entirely unknown. There is some evidence to suggest that the Nightingale (*Erithacus luscinia*) is extending its range northwards and westwards; and this species very curiously illustrates the potency of our Law of Dispersal, for there is not a scrap of evidence to suggest that its line of Emigration is spreading south of Exeter—a fact of very great significance, as we shall learn in a later chapter (*conf.* pp. 215-217). This species is extending its range higher in our area than elsewhere, due perhaps to the absence of the competing form *L. philomela*. The Whitethroat (*Sylvia cinerea*) appears to be increasing its range in the north of Scotland and becoming more dominant. The Willow Wren (*Phylloscopus trochilus*) is undoubtedly increasing its area, and also becoming more dominant in the north of Scotland; whilst the Wood Wren (*Phylloscopus sibilatrix*) is a yet more interesting instance, having spread northwards of late years even to areas at the very extremity of the mainland. In Ireland, too, although excessively local, a marked and recent northern extension of range is apparent. It is interesting to remark that the Wood Wren is absent from Norway, and there can be little doubt will never colonize the south of that country, even if it succeeds in penetrating to districts further north, say of lat. 60° (*conf.* footnote, p. 127). The individuals of this species that breed in Sweden and even as far north as Archangel winter probably in Abyssinia, hence its absence from Heligoland on passage—a line of Emigration to which we have already alluded at some length in the preceding chapter. The Marsh Warbler (*Acrocephalus palustris*) appears also to be spreading northwards in England, but as this species is so excessively local, and its distribution by no means perfectly known, it is perhaps wisest to discard this as an instance of recent emigration until we are in possession of fuller details. The Sedge Warbler (*Acrocephalus phragmitis*) has either not been observed in the Orkneys until I believe it was found by Mr. T. E. Buckley, or has extended its range to those islands recently. The Goldcrest (*Regulus cristatus*) is rapidly becoming more dominant and widely dispersed in Scotland, Gray stating that seventy years ago the bird was very scarce and local. It is also a significant fact that the great waves of migratory individuals of this species from the East in autumn, and which spread west into Ireland, normally strike our coasts south of Fife, thus indicating the area from which their ancestors emigrated east across the then North Sea Plains in past ages, and when the Goldcrest must have been a by no means dominant species in Scotland, although it certainly reached Scandinavia by way of the Orkneys and Shetlands. The Hedge Accentor (*Accentor modularis*) is unquestionably increasing its range northwards in Sutherlandshire and Caithness; and it may be remarked is another precisely similar instance to the preceding one of the Goldcrest—a species that has emigrated east from England across the North Sea Area before it extended its northern range at all dominantly into Scotland. The Nuthatch (*Sitta cæsia*) is one of our most local birds, yet there is evidence to suggest that it is steadily advancing northwest into Wales, where comparatively recently it was very rare. The Tree Pipit (*Anthus trivialis*) has, it is said, only been detected breeding in Sutherlandshire within the past twenty years.

The Greenfinch (*Fringilla chloris*) must also be regarded as another species that has extended its northern emigration in our islands within recent years. It has become much more dominant in the northern and wilder portions of our islands, as the cultivation of trees has increased, even reaching the Orkneys, where it now breeds sparingly. It is interesting to remark that this bird, like the Goldcrest, is only known as a Coasting Migrant over, or winter visitor to, the Shetlands; and this fact seems to me to suggest that the favourable conditions which assisted the emigration of this species across our area *viâ* the Orkneys and the Shetlands to Scandinavia (where it ranges up to lat. 65°), have lapsed—a variation in the climate of these islands involving the destruction of forest growth seems to have occurred (and there is some evidence that such actually took place), and by extermination driven the breeding range in our islands southwards for a time (although the area in question has still been crossed on passage by birds that have continued to breed in Scandinavia); but which, as we have seen, there is now absolute proof that it is slowly expanding north again with the return of more favourable conditions. This extension of area, it should be remarked, is being carried out by individuals that breed in Scotland, and is quite independent of the migratory individuals that cross it on passage to Scandinavia. The increase of the Starling in Scotland presents very similar facts.

The extension of range of the House Sparrow (*Passer domesticus*) has so very much depended upon the spread of cultivation and the reclamation of waste grounds, that it is almost impossible to separate the thoroughly normal increase of area of this species from that which has solely depended upon artificial aid. However, the species is one that is unquestionably extending its area of distribution, due in a great measure to its abnormal increase in numbers (owing to the extermination of many of its enemies and its artificial conditions of existence), and is now common and well established in many northern areas, where within the range of historic time it was absolutely unknown. The Tree Sparrow (*Passer montanus*) is also extending its range northwards, and apparently increasing in numbers; but it is most significant that this fact does not apply to the south-west of England—a direction of extension which would involve a southern emigratory movement. A few pairs of this Sparrow reached the Faroes some twenty-five years ago, and have multiplied to such an extent as literally to become a pest! The Chaffinch (*Fringilla cœlebs*), again, is increasing and spreading northwards in many directions in Scotland, but apparently has not extended its breeding area yet to the Orkneys, although it passes those islands and the Shetlands in autumn and spring, and is found in them in winter—an exactly analogous instance to that of the Greenfinch and Starling. The Bullfinch (*Pyrrhula vulgaris*) of late years has increased its range northwards to some of the Hebrides, especially to the south-eastern portions of Skye. The Starling (*Sturnus vulgaris*) is a specially noteworthy instance of current emigration, having within the past forty years extended its range north and west to an enormous extent. More especially has this been the case in Scotland, where within the past half-century it has spread into many districts where previously it had been quite unknown. Much of this range extension is probably due to the same causes as those which have spread the House Sparrow so widely during recent time.[1] The Jay (*Garrulus glandarius*), in spite of decreasing numbers due to incessant persecution, has extended its range northwards within a comparatively recent time, and now breeds at least as far north as Inverness-shire. The Magpie (*Pica caudata*) is said to be increasing its area of distribution in Ireland, although tradition states that it was originally introduced to the island by artificial

[1] For a very exhaustive account of the recent range expansion of this species in Scotland, I must refer my readers to Mr. Harvie-Brown's admirable paper in *The Annals of Scottish Natural History* (January 1895). I am inclined to think that the facts in many cases would bear a very different interpretation. This may be due perhaps to the distinguished naturalist who has collected them not being cognizant of what I believe to be is the Law of Life's Dispersal. Speaking generally, the phenomenon very closely accords with that Law of Dispersal as demonstrated in the present work.

means. Another very interesting instance of current emigration is furnished by the Rook (*Corvus frugilegus*). As usual, this line of extension is mostly northwards, and has gradually extended until the bird has reached and occupied the Orkneys, and possibly the Shetlands, as a breeding species. A north-western line of Emigration has also been followed, which at present extends as far as Skye, but there seems every possibility of a speedy settlement over the Outer Hebrides.

The Tawny Owl (*Strix aluco*) within the past half-century has extended its range northwards over the greater part of Scotland—including Skye and some other of the Inner Hebrides—in spite of its decrease in numbers. It still continues, however, significantly absent from Ireland, notwithstanding this emigratory movement elsewhere, and though common enough in the south of Scotland, it never attempts a southern dispersal in spring to breed in that island. The Ring Dove (*Columba palumbus*) is another very remarkable instance of current emigration. Increasing rapidly in numbers, it has initiated a northward extension of area on a very large scale, and within the past century has gradually spread northwards and westwards as a breeding species over Scotland, even to the Hebrides. This species is in its emigrations and migrations analogous to the Greenfinch and the Chaffinch. The Stock Dove (*Columba œnas*) in its recent emigrations presents some very interesting facts which confirm the soundness of our Law of dispersal in no uncertain way. Twelve years ago the Stock Dove was generally though locally distributed throughout England and Wales, and in the extreme north-east of Ireland, where, however, it was said to be

N

rare. In Scotland at that time but four instances of its occurrence had been recorded, viz. once in Stirlingshire, twice in Perthshire, and once as an abnormal migrant to the Orkneys. Now the bird has gradually increased its range northwards over Scotland, and is known to breed at least as far north as the shores of the Moray and Dornoch Firths! The occurrence of this species in Ireland in the north-east alone might appear almost like a southern emigration from Scotland, but fortunately we have positive evidence that the bird had not then appeared in the latter country. There is a strong probability that the range of this Dove in Ireland is not yet accurately determined. The Great Crested Grebe (*Podiceps cristatus*) has within quite recent years apparently extended its breeding range northwards into Scotland. The Woodcock (*Scolopax rusticula*) seems also to be extending its breeding range in our area, but this may be due to closer observation on the part of naturalists. The table opposite contains the species that are increasing their range in the British Islands.

This list of birds in the actual process of extending their range numbers twenty-nine species. Of these eight are Summer Migrants to our islands (indicated SM in the table opposite), and the remaining twenty-one are Residents, or species of which individuals may be found in our area throughout the year (indicated R in the table opposite). Dealing first with the summer migrants, we find that in all cases the birds that are so steadily increasing their northern range in our area did not emigrate across that area to Scandinavia, where, however, in no less than six instances out of the eight given their range extends much further north even in some cases far beyond the Arctic Circle to the North Cape, their Post-Glacial Emigration having been a

SPECIES.	RESIDENT OR SUMMER M.	IN NORTHERN BRITISH AREAS.	NORTHERN LIMITS IN WEST EUROPE.
Turdus viscivorus ...	R	Migratory.	lat. 66½°.
Turdus musicus	R	Migratory.	lat. 66½°.
Merula merula	R	Migratory.	lat. 67°.
Ruticilla phœnicurus ...	SM		lat. 71°.
Erithacus rubecula ...	R	Migratory.	lat. 66½°.
Erithacus luscinia ...	SM		South of Baltic.
Sylvia cinerea	SM		lat. 65°.
Phylloscopus trochilus ...	SM		lat. 70°.
Phylloscopus sibilatrix	SM		lat. 58°.
Acrocephalus palustris	SM		Denmark.
Acrocephalus phragmitis	SM		lat. 70°.
Regulus cristatus	R	Migratory.(?)	lat. 66½°.
Accentor modularis ...	R	Migratory.	lat. 70°.
Sitta cæsia	R	Sedentary.	Denmark.
Anthus arboreus	SM		lat. 69°.
Fringilla chloris	R	Migratory.	lat. 65°.
Passer domesticus ...	R	Sedentary.	lat. 66½°.
Passer montanus	R	Sedentary.	lat. 67°.
Fringilla cœlebs	R	Migratory.	lat. 70°.
Pyrrhula vulgaris	R	Sedentary.	South of Baltic.
Sturnus vulgaris	R	Migratory.	lat. 69°.
Garrulus glandarius ...	R	Sedentary.	lat. 66½°.
Pica caudata	R	Sedentary.	lat. 71°.
Corvus frugilegus	R	Sedentary.	lat. 66½°.
Strix aluco	R	Sedentary.	lat. 64°.
Columba palumbus ...	R	Sedentary.	lat. 66°.
Columba œnas	R	Sedentary.	lat. 61°.
Podiceps cristatus ...	R	Sedentary.	lat. 57°.
Scolopax rusticula ...	R	Sedentary.	lat. 66½°.

purely continental one to Scandinavia. Of the twenty-one resident species we find that in no less than fifteen cases the birds that are gradually emigrating northwards in our islands did not reach Scandinavia by a Post-Glacial Emigration across our area; whilst in the six instances of which we have evidence that such a line of Emigration did actually take place, some change of climate or of arboreal conditions has for a time lowered the breeding range in our area by extermination, and the lost ground is now slowly being recovered. With two exceptions all these resident birds range much higher in continental Europe than they do with us. The exceptions are the Nuthatch and the Bullfinch, represented however in more northern latitudes by closely allied climatic races. This fact, as applicable to the Summer Migrants as to the Residents, shows very clearly that it is only the indigenous or British individuals of the species that are endeavouring to expand, and in some cases to regain, their northern range, the dominant line of Emigration of the indigenous continental individuals being in the majority of cases entirely outside our limits, and, in the remaining few cases, principally so. We may also here remark that species in which the habit of Migration is dominant (although individuals may be sedentary with us or partly so) appear to extend their range more rapidly than strictly sedentary species—the migratory Ring Dove, for instance, advances quicker and more dominantly than the much more sedentary Stock Dove.

There is one very important fact connected with all this current emigration, and that is its invariable northward tendency. It may be in an easterly or westerly direction, but the trend is always north, never south. I find that in the majority of cases this emigratory movement has been attributed to the spread of cultivation, to the planting of trees and the making of shrubberies; but are we justified in accepting such an explanation, when we invariably find that this movement has been ever northerly, and of course in accordance with a known Law? Has cultivation and tree planting always been in a northerly direction? surely much of it has taken place in southerly areas as well; but with no accompanying signs of Avian Emigration from north to south! You may fill South Devonshire with reed-beds and with spinneys, suitable in every way for the Reed Warbler and the Nightingale, yet the birds will not emigrate southwards to fill them; you may afforest Ireland as densely as you like, as indeed much of it has been afforested, yet you cannot tempt a single Woodpecker or a single Tawny Owl or Nuthatch to break the Law of its dispersal or the conditions of its extension, either by emigrating south or by crossing a wide water area to take up its abode in your forests. I do not deny that this afforestation offers many advantages of which species have readily availed themselves and increased their area, but only is such an advantage seized upon when it is also in harmony with an unvarying Law of dispersal. Depend upon it, it is the same Law of dispersal that governs the emigrations of species in our islands now, just as it governed the movements of species in other areas, and as we know it controlled their extension in past ages. Not a single instance of a species increasing its range in our area in a southerly direction, or from north to south, can be named among the long list of examples given above. Of course we except instances of dispersal which may have an apparent southern tendency in such species that have been introduced into certain areas by artificial means—the Red-legged Partridge, the Capercaillie, and some others, that have been brought to the British Islands (and maintained there) by human agency.

There is another very interesting fact which must be noticed in connection with this extension of range of so many species in the British Islands. In no less than seventeen cases out of the twenty-eight instances of which particulars have been given has this Emigration been attended by Migration—the latter has helped the former; in fact it is doubtful whether a successful extension could have been established at all had the species remained non-migratory, or had not already been addicted to migratory habits, for the winter conditions would have been fatal to the colonists, and the range gained in summer would have been lost in the following winter. The sedentary species must therefore have increased their area more slowly, not extending their range north into districts until the winter conditions were favourable. The Migration taking place within our islands is precisely the same as that Migration which progresses beyond them; the flight south of a Song Thrush in autumn from the north of Scotland to more southerly areas in the same country is precisely the same movement as the flight of a Knot from Grinnell Land to South Africa; the difference is only one of degree. One can readily understand how individual birds and their offspring gradually extend their range year by year, spring by spring resorting to a locality to breed, yet compelled by the severity of the ensuing winter to return to their more southern base—as the range expands northwards the migration flight lengthens, until instead perhaps of a few miles only the journey slowly becomes one of many miles. The route followed is the line of extension in which the breeding range has been increased, each individual bird following a road back in autumn which it traversed north in spring—a route which has been slowly acquired by the species mile by mile, even field by field, or wood by wood, and taught to the young that journey south in company with the old birds in autumn, or that follow them north again in spring. So far as the Summer Visitors that are extending their range in our islands are concerned, the movement is slowly increasing the length of their journey, and slightly varying the route that they follow; with the Resident species it is in many cases compelling the adoption of local migratory habits. It may be remarked that this local migration is much more marked and general on the Continent than it is in our island area, where the climate is so much milder; more species perhaps partake in it, and the flights are longer and more dominant, and rise to a much greater degree of importance. Thus we find many species absolutely migratory in continental areas, no individuals remaining through the winter that are sedentary or nearly so with us, in the sense of never being absent from our islands all the winter through. In the above table I have indicated the species that are migratory in the northern portion of their British area, but we need not stay to enter into greater details of their movements. Some of these, I may remark, have already been described in the *Migration of Birds*, and to that volume I would refer the reader who may be sufficiently interested in the subject to desire fuller information.

In order to render the present subject fairly complete it is now necessary to devote a little space to the consideration of a very different set of facts from those which we have been studying; to make a brief allusion to those species which have passed more or less completely from our avifauna, and whose absence has been very largely caused by those very improvements that have been so advantageous to other species. Our drainage and reclamation of waste lands, our game preserving and mania for collecting "British" specimens, together with our rapid increase of population, have cost us dear, so far as some of our most interesting birds are concerned. Many of these species formerly resident in, or regular migrants to, our islands have been completely exterminated, or only survive in the northern thinly populated and least cultivated districts. As may naturally be surmised, nearly all the smaller birds have maintained their ground, even in many cases increased their range and their numbers, but it is at the expense of larger and perhaps more interesting birds. The species in question are named below.

* Savis Warbler.	Kite.	* Crane.
Bearded Titmouse.	Honey Buzzard.	* Great Bustard.
Short-eared Owl.	† Osprey.	Dotterel.
Marsh Harrier.	* Bittern.	* Avocet.
Montagu's Harrier.	* Spoonbill.	† Ruff.
† Golden Eagle.	† Gray-Lag Goose.	* Black-tailed Godwit.
† White-tailed Eagle.	Capercaillie.	* Black Tern.

No less than eight of these (marked by an asterisk) have become absolutely extinct as breeding species in the British Islands; five others (marked by an obelisk) have been banished from England and now only breed in Scotland; two of them, however, nesting in Ireland. The remaining eight are very locally distributed, but still continue to breed in our islands, although there can be little doubt that most of them will eventually disappear if some special means for protecting them are not quickly devised. Every naturalist must deplore the absence of the Bittern, the Spoonbill, the Crane, the Great Bustard and other interesting birds from our avifauna; all the more so, for in perhaps every instance by a little judicious management the species might have been preserved to us. It is too late now in many cases for us ever to hope to see these beautiful birds re-established in their ancient homes, or in what remnant of their haunts still remains untouched or unimproved by modern methods. Could we turn half England once more into a vast morass, the Laws of dispersal would forbid the Bustard, the Spoonbill, and the Wild Goose to enter. As colonists they will never return again, for now our islands are beyond the limits of their normal migrations, and our wide water areas, so long as they continue, will ever act as a barrier to successful re-emigration. The Golden Eagle, the White-tailed Eagle, the Dotterel, and the Ruff are the only species whose lines of migration now cross our islands at all dominantly, and they are the only species that are ever likely to re-people the areas from which they have been banished, although even this is very problematical, notwithstanding any and every inducement we might offer them to do so. The Short-eared Owl, the Marsh Harrier, Montagu's Harrier, the Kite, the Honey Buzzard, the Osprey, the Ruff, the Black-tailed Godwit, and the Black Tern we might yet save, were strict protection given them; but the few individuals that are from time to time destroyed at their old breeding grounds are assuredly among the last of their kind that will ever visit us with the object of reproducing their species, and if we still continue to kill them or to drive them thoughtlessly away, the opportunity of retaining them in our land will soon be gone for ever. They are the last remnants of the individuals whose lines of emigration extended to our area, and taking into consideration our insular position, they are the last that will ever essay the attempt. Nothing, then, short of absolute re-introduction by human aid will re-establish them in our islands, as we know by experience. Such sedentary species as the Bearded Titmouse will never again attempt to enter our area normally. The Capercaillie, also a sedentary species, would never normally have become a British species again had not man's assistance been given, although we know for an absolute fact that our islands were and are suited in every way to its requirements, as experience has shown. Nothing short of re-introduction by man would ever establish the Gray-Lag Goose in the lowland counties again; as a breeding species it has passed north never normally to return. Professor Newton, in his very interesting paper on the Great Flood in the Fens during the winter of 1852-3 (*Trans. Norfolk and Norwich Nat. Soc.*), records that among other species that appeared with the return of more favourable conditions, due to the flooding of the land, were the Black Tern and the Redshank, which might seem like instances of pure re-colonization; but both these species have not yet been entirely banished from the district. No return of such utterly banished birds as Spoonbills, Avocets, Cranes, and Gray Lag Geese (all completely exterminated) was remarked.

Knowing then what we do of the Laws of Avian dispersal, let us employ our knowledge while there is yet time in saving at least a remnant of that rich and interesting avifauna that has vanished from our shores, by protecting that remnant jealously from persecution and wanton, senseless destruction.

CHAPTER VI.

ISLAND AVIFAUNAS.

The West Palæarctic Islands—Continental Islands—Birds of Borneo, Formosa, the Philippines, etc.—Ancient Continental Islands—Birds of the Canaries, Madagascar, Azores, Bermuda, etc.—Reasons for the Unequal Dispersion of Species—Islands and Migration—The British Islands—Endemic British Species—The Red Grouse—Endemic British Races, or Representative Forms—The St. Kilda Wren—Races of Titmice—Poorness of British Avifauna in Endemic Species—The Channel Islands and Heligoland—West Mediterranean Islands—The Canary Islands—Endemic birds of—Number of Eggs laid by Birds in Canary Islands—Madeira and the Azores—Japan and the Bonin Isles—Various Tropical Islands—Endemic Avifaunas of—Bearing of Migration on Insular Avifaunas—Conclusions Drawn from Facts—Bearing of Glacial Conditions on Island Avifaunas.

WE have now reached that stage in our investigations where it becomes necessary to deal more specially with the avifaunas of islands, and to ascertain what relation the birds of the various islets that dot the western limits of the Palæarctic Region bear to the species that inhabit the large land masses or continental areas adjacent to such island groups, and the bearing of that fact upon our whole subject. We may commence by stating that all the west Palæarctic islands, with two exceptions, are what are termed Continental Islands, or islands that at some more or less remote epoch formed part of the continent to which they are adjacent. The two exceptions are the Azores and the Madeira groups, which are oceanic islands, and apparently of volcanic or coralline origin. The continental islands resolve themselves again into two distinct classes, viz. those that are of ancient origin and usually separated from the parent land mass by deep seas; and those that are of recent origin, always situated upon submerged banks that connect them with the adjoining continent, and surrounded by shallow seas. The Canary Islands may be taken as the only example of ancient continental islands in the West European Area; whilst the British Isles, the Channel Isles, Heligoland, the Balearic Isles, Corsica, Sardinia, Sicily, and Malta all belong to the class of recent continental islands. The more ancient islands very often contain the greater number of peculiar species, due to their longer isolation, and are remarkable for the fragmentary character of their fauna. The more recent islands are said to possess few peculiar species, and to exhibit the general characteristics of the fauna and flora of the adjacent land mass. The volcanic or coralline islands are probably peopled entirely by fortuitous emigration, by accident, and therefore often exhibit a somewhat puzzling mixture of species.

In the present chapter it will be my principal aim to show why these Atlantic and Mediterranean islands are, and in all probability will continue to be, so poor in peculiar species, and why various other islands in other parts of the world, irrespective of their age and origin, are respectively rich or poor in endemic species—a question which has hitherto, so far as I can determine, never been grappled from the same standpoint as that which

I propose to take. Dealing first with continental islands of recent origin, why, we may very naturally ask, does Borneo—presumed to be a recent continental island—contain so many peculiar birds, whilst the British Islands, of probably the same geological age, contain so few endemic species? Why does the Japanese Archipelago exhibit such a paucity of peculiar forms, whilst Formosa—only about half the size of Ireland—is rich to an astonishing degree in endemic species? Why should Ireland possess not a single endemic bird, whilst the Philippines are replete with them?

Of the 580 species of birds that have hitherto been found in Borneo, according to the latest available information, no less than 108 species are peculiar to the island. In Formosa the late Mr. Swinhoe met with some 144 species of birds; subsequent investigation has shown that no less than 43 species are endemic; whilst of the 472 species that have been obtained in the Philippine Islands, no less than 300 species are peculiar! On the other hand, the list of birds found normally in the British Islands contains, say, close upon 250 species, but out of this number only five are endemic, and four of these cannot even claim specific rank. Again, the list of birds found in the Japanese Empire contains some 381 species (*fide* Seebohm), but only seventeen are known to be endemic, and of these five at least are not specifically distinct. The Channel Islands do not contain a single endemic species; Corsica but one (*Sitta whiteheadi*); none, so far as is known, occur in the Balearic Isles, in Sardinia, Sicily, Malta, or Heligoland. Of the two latter islands it is interesting to remark that no less than 278 species were recorded by Mr. Wright from

Malta; whilst the more astonishing total of 396 has been recorded from tiny Heligoland[1]—yet neither island can claim a single endemic species or race!

Turning next to continental islands of more ancient origin, we find that in the Canaries (the only group which can claim such a distinction in the West Palæarctic Area) the number of species recorded by Mr. Meade Waldo is 146, of which perhaps ten are endemic, but most if not all of these are only island forms or representative races of continental Palæarctic species. On the other hand, Madagascar, another continental island of ancient origin, contains 238 species, of which no less than 129 species or races are endemic. Not a single genus is peculiar to the Canary Islands, but no less than 35 genera are confined exclusively to Madagascar! As an example of Oceanic Islands we can compare the Azores and Madeira with say the Bermudas and the Galapagos. On the Azores some 53 species of birds have been obtained, 38 of which are residents, and 15 only stragglers to the group: two species are endemic, one of which is very closely allied to Canarian and Madeiran forms. Madeira numbers some 100 species, either as residents or abnormal migrants, and can claim but two endemic birds, one of which at least is only subspecifically distinct from a Canarian form. Coming now to the Bermudas, we find that no fewer than 180 species have been observed on the islands, but

[1] This is the number recorded by Gätke, but at least one species has been shown to have been so in error—*Geocichla dauma* (conf. *Ibis*, 1894, p. 298). Probably some other records are equally unsatisfactory, so that the total number of Heligoland species may not be quite so large.

only 11 are resident, the remainder being abnormal migrants of more or less frequency. Not a single species is endemic. The Galapagos can claim 57 species of birds, and of these no less than 38 are absolutely endemic, all the land birds (31 in number) being peculiar except one species, and more than half generically so!

The facts briefly stated above are very interesting ones, and apparently inexplicable by any known laws of avian dispersal. The anomalies, however, are but apparent, the difficulties of explanation are far more imaginary than real. This very large amount of apparently anomalous and unequal dispersion of species over these island areas is probably entirely due to geographical causes, correlated with avian migration. I think it may be safely laid down as a universal rule that islands situated on a direct and important line of Migration are never remarkable for a great or predominant number of endemic species, or even of local races, and it is the exception to find any such peculiar species at all. On the other hand, islands remote from any dominant line of Migration just as invariably contain proportionately or absolutely large numbers of endemic species or peculiar races. The islands most remarkable for their number of endemic species are almost without exception situated in the Southern Hemisphere, or within the Tropics, where Migration does not prevail to anything like the extent that it reaches in the Northern Hemisphere, and where—as in the Tropics—the climate is one of uniform warmth and stability, rendering migratory movement on any very important scale unnecessary. Briefly, then, we find the great number of endemic or peculiar species where the avifaunas, both on the islands and in the surrounding areas, are sedentary. Let us deal with each of these island groups in turn, and see how the facts they present may be explained by geographical position, and the absence or presence of any dominant line of Migration.

First we will consider the British Isles. We have already dwelt at some length on the emigration of species to this area, and on the comparison of its avifauna with that of adjacent areas. For the purposes of the present inquiry it will only be necessary to confine our attention to the few endemic species or races that dwell in these islands, to explain their presence in that area, and to show why the British avifauna is so poor in peculiar forms. At the close of Chapter IV. (p. 158) we gave a list of these British endemics. But one of the five species is absolutely distinct, the remaining four being races or insular forms, only subspecifically distinct from continental species. The one exception is so utterly unique in character that I cannot resist the opportunity of inquiring somewhat fully into its origin. The Red Grouse is absolutely peculiar to the British Archipelago, the one solitary endemic bird entitled to specific rank. How did it come there? The nearest ally of the Red Grouse, in fact its continental representative, is the Willow Grouse (*Lagopus albus*). There can be no reasonable doubt that before the Glacial Epoch the Willow Grouse was the only and dominant species in West Europe, including what is now the British Area. Exterminated in the north by the changing climate, by the snow-fields and the glaciers, the remainder of the species continued to dwell in South

Europe, and the evidence strongly suggests that the range of this Grouse reached far to the south-east, from Scandinavia and North Russia as far as the Caucasus (where it is probable this Grouse will yet be discovered as an indigenous species; at least one example has been obtained there), Asia Minor, and Turkestan. Owing to its arboreal habits, it soon became exterminated in the north, its disappearance following that of the willows and birches. The birds inhabiting what was then the British portion of continental Europe were likewise exterminated in the north by the glacial climate, say to limits that reached from 53° north latitude, down to the Pyrenees. In the course of time the great glaciers and snow-fields of Central Europe completely isolated the two colonies of Grouse. Those in the south-west were probably never exposed to such a severe climate as the individuals dwelling in the south-east. Be that as it may, there can be no doubt that the Willow Grouse that occupied that portion of the Pre-Glacial range of the species composing Refuge Area I. during the Ice Age were the ancestors of the Red Grouse, and must have been differentiated to a certain extent by their isolation at that period. During their long sojourn in this area, in their probably severe struggle to maintain themselves, they became heath or tundra birds, compelled to give up their arboreal habits; and as the climate ameliorated and their descendants gradually spread north again, they would naturally keep to those districts that presented the least changed conditions—the mountainous areas, the watersheds and moors—as they still continue to do, not emigrating across the low fertile plains of the North Sea Area, and perhaps finding no country suited to their requirements from North France to Denmark, those areas being entirely non-mountainous, flat, and probably of a dense forest or swampy character. That the dominant line of Post-Glacial Emigration northwards of this party of Grouse (whether completely differentiated then or not being a matter of no consequence) was only across Britain, and did not extend in that direction to Scandinavia, is proved by the absence of this bird from the Shetlands. That it emigrated across our area slowly seems also suggested by the fact that no trace of an emigration in the direction of Greenland by way of the Faroes is to be found. Probably by the time this Grouse had reached the Orkneys, this north-westerly route was considerably broken by areas of water,[1] or conditions favourable to range extension in this direction were wanting. The Willow Grouse that now dwell in Scandinavia have reached that area by a north-westerly line of emigration, and there is no trace that this extension was ever west of E. long. 10°. The almost sedentary habits of these Grouse have been the means of keeping the two colonies distinct, and by isolating the individuals to allow of the differences—due to changed conditions of life—to become constant characters of full specific value. The brown dress of the Red

[1] Even had the Red Grouse reached Scandinavia by an emigration across the British Area (of which there is no evidence) or by way of Holland and Denmark (which is by no means improbable), the species must have been exterminated again by the fourth Glacial Period—the epoch of the Great Baltic Glacier—which, it will be remembered, only slightly affected the British Area, not sufficiently to cause the extinction of such a hardy tundra species as the Red Grouse.

Grouse was retained in winter owing to the pluvial climate of the mild western region—a change which the bird has been quick to take advantage of as a means of protecting itself from enemies amongst its native heath and ling. The propensity of the Red Grouse to perch occasionally in trees is very interesting, and unquestionably a relic of the arboreal habits of its ancestors. The emigrations of the Red and Willow Grouse, it will be remarked, are almost exactly analogous to those of the Eastern and Western Nightingales.

Of the remaining four endemic birds, the St. Kilda Wren (if it can be preserved from extermination) has probably the best opportunity of asserting finally its specific distinctness from the Wren that inhabits the rest of the British Area. It is, however, wrong to presume that St. Kilda owes its Wren to a party of birds driven south from Norway in search of a milder winter climate. In the first place no species, nor single individual of a species, normally migrates south along an unknown route which its Post-Glacial emigrations north have not followed. No bird, no species, emigrates or extends its area in winter, colonizing movements are prompted solely by reproduction.[1] Of course it may be urged

[1] One most convincing proof that species do not increase their area during winter, or retreat by emigration from adverse conditions, is furnished by the fact that great numbers of species, or portions of species, resort to winter quarters much further south or more remote from their breeding grounds than there is any necessity so to do. To my mind this is proof beyond the possibility of doubt that these remote winter areas are ancient range-bases from which the species has been dispersed, and from which it has colonized the most remote parts of its present area. The Willow Wren, for instance, goes in some numbers as far south as The Cape to winter, but many individuals remain in North Africa and even in South

that the movement to St. Kilda was perchance a purely fortuitous one, but the possibility of that being the case is destroyed at once by the fact that St. Kilda was completely glaciated during the Ice Age, and every bird exterminated from its surface. The Wren reached St. Kilda by the line of Emigration which we have seen extended from the British Isles to Greenland, and by which same route the Faroe Isles and Iceland received their *Troglodytes borealis.* That both these insular Wrens preserve their subspecific identity, and (if the present conditions continue) may ultimately attain complete specific rank, is entirely due to the fact that Migration along this route—so far as these subspecies are concerned—has ceased; the Wren (in its various subspecific forms) is now sedentary, and isolation is therefore complete. The three subspecies of Titmice that inhabit the British Area, however, may very probably never rank higher than they are to-day, because the migration of the continental (and parent ?) species continues to pass our area. So long as this migration and consequent intermixture of individuals continue, the swamping effects of intercrossing will prevent complete segregation; were the migration, however, to lapse, isolation would doubtless soon enable

Europe at that season. Numbers of Chiffchaffs go as far south as Abyssinia to winter, whilst others spend that season even in the South of France, Iberia, and Italy. Many Sedge Warblers winter in Algeria and various parts of North Africa, others journey thousands of miles further to the south in that continent. It is nonsense to suggest that these birds are seeking a suitable winter climate so far to the south, as many individuals of the same species find suitable winter conditions at a distance measured by thousands of miles nearer to their breeding grounds. Scores of similar instances might be given.

complete distinction to be attained. I cannot admit what Dr. Wallace asserts to be the case, viz. that in the British Isles "the process of formation of peculiar species has only just commenced" (*Island Life*, new ed., p. 408), for there is no evidence whatever to show that these British endemic forms are not of very ancient origin—dating, it may be, from the close of the Pleistocene Period—to indicate that they are not struggling with adverse conditions in segregating themselves; conditions, be it remarked, that will prevent complete specific distinction so long as they continue.

Now a few words with regard to the poorness of the British avifauna in endemic species. The geographical position of our islands is directly opposed to the establishment of local or peculiar species. They are situated too far north for many species ever to become absolutely sedentary within their limits (so long as climatal conditions continue the same), and they are adjacent to areas where the conditions for sedentary residence are even less suitable. The inevitable consequence is that the several avifaunas are constantly being mixed by the migration taking place over these various districts; and to such an extent does this prevail, that probably not more than two or perhaps three of our indigenous birds are sufficiently isolated from the remainder of the species to ensure the preservation of any variation that they might develop. The species that are the most sedentary show the greatest amount of local variation. This is specially observable in the case of the Dipper, the Red Grouse, and the Partridge; and could we only isolate sufficiently the various local types of these species that occur in certain areas, there can be little doubt that a subspecific and eventually a specific distinction would result. We may conclude, therefore, beyond the slightest doubt, that our paucity of endemic species is due to the vast amount of migration taking place over our islands; and so long as the migration of any and every species continues, so long shall we remain poor in peculiar forms. Of plants, insects, fish, and animals, however, whose means of dispersal are more limited, and whose habits are sedentary, we have a fair proportion of endemic species—a fact which only emphasizes the truth of the foregoing remarks.

Let us now pass to the Channel Islands and Heligoland. So far as the Channel Islands are concerned precisely the same remarks apply—their geographical position preventing the establishment of a single endemic species. Their avifauna contains a fair proportion of species, composed of residents, summer migrants, winter visitors, and coasting migrants; yet none of the resident species are isolated from the vast stream of migratory individuals of the same species regularly crossing their area. All the species indigenous to them are continental or British, and none of them are absolutely sedentary. Heligoland, situated about 40 miles from the mouth of the Elbe, and only about one-fifth of a square mile in area, is said (on the authority of Herr Gätke) to have been visited by no less than 396 species,[1] yet not a single endemic bird is found thereon—a fact entirely due to the abundance of migratory birds that so regularly pass over it.

We will now proceed to notice the various West

[1] See footnote, p. 190.

Palæarctic islands in the Mediterranean—the Balearic Islands, Corsica, Sardinia, Sicily, and Malta. It may be remarked that all these islands are situated in a comparatively small and entirely land-locked sea, and that all of them formed part of continental land at no very remote era. So far as our knowledge extends at present, only one of this important series of islands contains an endemic species. In Corsica we have a species of Nuthatch (*Sitta whiteheadi*) peculiar to the island, so far as is known. That endemic species should be so excessively rare in islands apparently suitable in every way for the development of insular forms would appear to be a somewhat anomalous fact, but the explanation is an extremely simple one. Again, geographical position is the cause of the absence of endemic species; and each one of these islands is situated in the direct path of a strong migration, indeed acts as a sea-bridge to its progress. The resident avifaunas of these several islands are composed of species that range well into Europe, cross them on their way to Africa, or reside in them during the winter months. This intermixture of individuals—due to migration—has prevented any local races becoming established in these islands, with the sole exception of the Corsican Nuthatch. This exception, however, is one that proves the rule. The Common Nuthatch appears everywhere to be a sedentary species, and to be absent from Malta, Sardinia, and Corsica. The descendants of the individuals which in emigrating north across the Mediterranean after the Glacial Epoch became stranded residents in Corsica (and it is probable that *S. whiteheadi* occurs also in Sardinia), have remained sufficiently isolated to become segregated into a well-marked species. It is very probable that the Nuthatch of Algeria will be found to differ from the European type. It is therefore most significant that the only instance of an endemic species in these Mediterranean migration-swept islands is that of one whose parent form is non-migratory, and thus never interfered (by the intermixture of individuals) with the process of differentiation.

Passing now to the Canary Islands, we are confronted with a very different series of facts. As we have already seen, these islands in remote ages probably formed the extreme south-western termination of what was then continental Europe. They formed a part of a great Range Base of species during the Glacial Epoch, of which many traces still endure, either in the form of endemic races or species, or in the annual passage to and from them of many migratory birds. Their situation, however, at the very south-western limits or extremity of a refuge area, or range base on the northern limits of a wide expanse of water across which we have no evidence that purely terrestrial species penetrated, is a most favourable one for the establishment of endemic forms. The Canary Islands still continue to be the resort of a great many northern species in winter, but there is no evidence of any strong migration across them from north to south, for reasons which have already been dwelt upon. This group of islands within the past few years has received considerable attention from ornithologists, among whom may be mentioned Canon Tristram, Mr. Meade Waldo, and Herr Konig. The result of these investigations has been to show that the Canary Islands are, comparatively speaking, singularly rich in endemic forms; in fact no other group of islands in the West Palæarctic region can boast so many peculiar races or such abundant examples of insular variation. The explanation is a very simple one, and the facts confirm in no uncertain way the truth of our general arguments respecting island avifaunas. The number of endemic races may be taken at ten, although this does not quite represent the peculiarities of the islands, for several species and races not included in that total are apparently common to Madeira and the Azores as well. Of these ten no less than three are island forms of two endemic species, confined to various parts of the Canarian Archipelago, some of which apparently interbreed among themselves. How have these endemics become differentiated, and how do they continue to preserve their characteristics? It will be remarked that not one of them is an endemic race of a species that now visits the islands on migration, or at least that portion of the archipelago inhabited by the peculiar form. They are all allied either to North-west African sedentary species, or to birds that breed in Europe and North-west Africa, yet do not extend their winter migrations to the islands. When the Canaries formed one unbroken area with the Atlas, there can be no reasonable doubt that these endemic species were not in existence. Being sedentary throughout this area, they have—as submergence or volcanic action gradually or suddenly reduced it to an archipelago—continued to inhabit the localities where they were so isolated, and have thus been able to preserve the differences which now entitle them to specific or subspecific rank. It is interesting to remark that the Canarian form of the

Robin (*Erithacus superba*) is confined to Teneriffe and Grand Canary, two of the most southerly and remote of the islands, where the ordinary form of the Robin is never seen; and that the individuals of *Parus ultramarinus* (a sedentary species ranging from Algeria to the Canaries) present some differences of size and colour. I may also remark that the Stonechats resident in Algeria may yet be found to belong to the Canarian form, or to resemble it more nearly than the migratory individuals of this species. Whilst dealing with the birds of the Canaries, I may take the opportunity of making a few remarks on the number of eggs laid by various species in the islands. It is a most curious and interesting fact that many birds breeding in the islands lay fewer eggs for a sitting than they are known to do elsewhere. Thus we have the Blackbird with clutches of two or three, frequently one; the Robin two or three; the Canarian Firecrest three to five; the Tits three to five; the Teydean Chaffinch two; the Barn Owl two; Bolle's Pigeon one; and the Canarian Pigeon one! This small number to the brood, so general among so many different species (especially in the case of the Pigeons), is possibly caused by the scarcity of food for the nestlings, or difficulty in obtaining it and conveying it to the nest—the result of a long process of natural selection.

With Madeira and the Azores we come to a slightly different class of phenomena. These islands are very remote from continental land, and their avifaunas are perhaps entirely derived from fortuitous emigration. Although Madeira numbers some 100 species, and the Azores 38, each of these island groups can claim but two endemic forms. No wonder need be expressed at the paucity of endemic species in Madeira and the Azores, for they are situated too near the vast migration that passes the coasts of West Europe, and are therefore constantly being visited by abnormal migrants, out of their proper course, which prevent the complete isolation of the island individuals, and destroy by interbreeding any tendency towards specific distinctness. I need scarcely remark that the endemic races are more closely allied to species that rarely, if ever, reach the islands. The various connections between the avifaunas of the Canaries, Madeira, and the Azores are certainly remarkable, and in some cases scarcely the result of fortuitous migration. The facts seem to suggest that these areas were formerly, and at no very remote era, more extended and in closer proximity than is now the case.

I might also remark that Japan, with its few endemic forms, is precisely analogous to the British Islands. It is situated on or near the direct path of migration that flows down the extreme east of Asia, and its avifauna is not sufficiently isolated ever to assume a very endemic character. It is significant that the Bonin Isles, out of the line of this dominant stream of continental migration, contain comparatively a very large number of endemic forms.

A brief notice of Tropical Islands, remarkable for their wealth of endemic species, now becomes necessary. For instance, we have Madagascar with its 129 endemic species, out of a total avifauna of 238; Borneo with 108 out of 580; the Philippines with 300 out of 472! Many scores of similar instances might be given, especially from the Malay and Pacific Islands. We need not inquire how or whence the avifaunas of these various islands have been derived; all that concerns us is by what means such a very large percentage of the species has succeeded in becoming endemic. The indigenous species of all these areas are sedentary. Vast numbers of birds may visit these islands in winter, or pass over them on migration, yet none of such birds are the direct ancestors of these endemic species. The endemic avifauna stands quite apart; its affinities are exclusively with the non-migratory species in the more or less adjacent areas, or with forms that from a variety of causes have ceased to visit the area occupied by the peculiar form. It is said that many of these endemic island species owe their origin to volcanic or seismic agency, a species often being isolated into various groups, as a previously compact area has been divided into various portions by earthquakes or volcanic action. This is true, but only under certain conditions. Japan is a great centre of earth disturbance, so are various parts of the Mediterranean; but we do not find the same results following upon such earth action as we find in other areas, as, for instance, in the Pacific—where almost every island possesses some endemic form—or in the East Indian Archipelago, where the phenomenon of peculiar species is very apparent and dominant. If the geographical area be situated upon a migration route, little specific change will result, no matter how varied the physical disturbance; if, on the other hand, the disturbance takes place (even on a comparatively small scale) in areas where no migration occurs among the avifauna, or where species are chiefly sedentary, it will lead to the establishment of great numbers of peculiar forms, proportionate to the amount of isolation produced by such physical change.

I think, from a careful study of the facts, that we are perfectly justified in coming to the following conclusions. Firstly, it is only in the Southern Hemisphere or within the Tropics, where migration is not very extensive or is entirely absent, that we find islands remarkable for endemic avifaunas; all the islands in the Northern Hemisphere probably being Post-Glacial and recent, so far as their avifaunas are concerned, those in the Southern Hemisphere being generally of greater antiquity, and dating to a very large extent faunally from the last glacial epoch at the South Pole. Secondly, that endemic *species* may be established only in such areas where the migration of the parent form has entirely lapsed: endemic *forms* or *races*, not specifically distinct, exist only in such areas where the migration of the parent form is only slight, not sufficient entirely to swamp the differences by interbreeding; and that such endemic races or forms are not necessarily evidence of species just commencing their segregation, for obviously they may be of considerable antiquity. Thirdly, endemic species are never closely allied to species that pass their area on migration. Fourthly, migration tends largely to preserve the ornithological identity of all areas over which it is dominant. Fifthly, changed conditions of environment are to a very large extent powerless to produce specific change if the geographical position is unfavourable to complete isolation from a dominant line of migration of the species affected.

The bearing of glacial conditions on the problem of

island avifaunas is a very important one; and it is probably entirely due to the influence of that recent Ice Age that the Northern Hemisphere is so utterly poor in endemic island species, it is a region of climatic forms; whilst on the other hand the rarity and shortness of migratory movement in the Southern Hemisphere, due to the present high state of glaciation of South Polar latitudes, together with the greater antiquity of existing conditions, to some extent explain the abundance of endemic forms in all isolated areas, from New Zealand to the limits of the northern tropic.

PART II.—MIGRATION

A STUDY OF THE PHENOMENON OF AVIAN SEASON FLIGHT ACROSS THE BRITISH ARCHIPELAGO.

MIGRATION ROUTE CHART

Migration of British Birds. Plate VI.

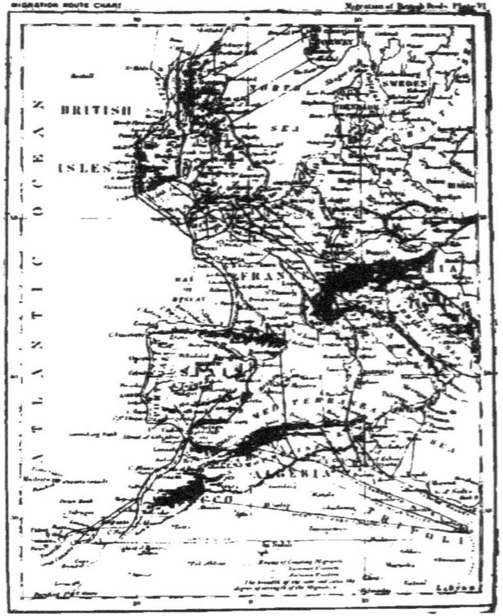

CHAPMAN & HALL, LIMITED

PART II.—MIGRATION.

CHAPTER VII.

ROUTES OF MIGRATION.

Difficulty of tracing Routes to British Islands—Definition of a Migration Route—The Gradual Effects of a Changing Climate on Birds—Impulses to Emigration and Migration—The Turnstone and the Rose-coloured Pastor—Ancient Breeding Ranges—Inter-polar and Inter-hemisphere Species—Breeding Grounds and Winter Quarters coalescing—Routes followed by Summer Migrants to British Area—Routes into the South of England—Species following them—Routes into Ireland—How followed by Birds—Absence of Routes into Scotland *viâ* Ireland—Past Physical Changes indicated by Present Routes of Migration—Persistency shown by Birds in following Migration Routes—Across the English and St. Georges Channels and the North Sea—Palmén's "Fly Lines"—The North Sea Routes—Origin of—Effects of Submergence on the Emigration and Migration of Birds—West to East Migration—Water Areas a Check to Emigration—Routes followed by Winter Visitors to and Coasting Migrants over the British Islands—The Routes of Migration that are most followed—Inland Continuation of Migration Routes—Difficulty of Tracing—Correlation of Routes with Breeding Grounds.

It is a somewhat difficult matter to trace the various routes by which migratory birds enter the British Islands, especially in the south. This difficulty is entirely due to our want of requisite information. I yield to no one in my readiness to admit the value of the work accomplished by the British Association, respecting the information collected by its committee concerning Migration in the British Archipelago. But one of the most important points of observation has been entirely neglected. I allude to the complete absence of information respecting migration between the Start Lighthouse in Devonshire and the Varne Light-vessel at the mouth of the Strait of Dover—the most interesting stretch of coast-line throughout the entire British Archipelago. Concerning the internal routes followed by migratory birds, our information is still more meagre. We now want a thousand recording stations in inland districts, with observations extending at least over a period of five years before we can obtain sufficient material to suggest very minutely the probable lines of migration within the British Area. This need be no difficult task, all that is required being the earnest co-operation of ornithologists. In the present chapter, therefore, I cannot treat the subject so fully as I desired; but I think we have sufficient material in our possession to suggest some very important facts.

In the first place, it may serve to simplify matters if we endeavour to explain What a Route of Migration is, and How it has been formed. The following remarks, be it clearly understood, apply as much to British species as to species in all other parts of the world; and the facts set forth must be constantly borne in mind by the student of Avian Dispersal and Migration. In the first place, then, we may remark that a bird's breeding grounds and its winter quarters were once continuous, no matter how remote one may now be from the other. The Northern Emigration has slowly progressed purely by an extension of breeding area, which in many cases has now ceased to be a breeding area in the extreme south, owing to a change in the climate, due probably to equinoctial precession; but in many more cases the breeding range and the winter range do still continue to overlap at the southern extremity of the one and the northern extremity of the other. The birds that breed exclusively in the high north required a Polar temperature, and were among the first to cease breeding in temperate latitudes; but there is perfectly conclusive evidence to show that when more southerly areas were Arctic in climate, the breeding range extended over them, as witness so many Polar birds wintering, if in small numbers, in comparatively high latitudes, as, for instance, the Knot in the British Isles.

It is only too popularly believed that the Glacial Epoch was a *sudden* phenomenon; that birds left the Arctic regions at once, as if overtaken by a change of climate as quick as that which now marks the change from summer to winter in the Arctic regions; and that parties of a species, or the entire species, immediately emigrated this way and that, along that coast-line or down that valley, far south to a refuge from its rigours. No more erroneous view of the facts could be conceived. Ages most probably elapsed before any perceptible difference could be observable in the northern range of species, and no one generation of birds would experience any perceptible change in the climatic conditions. As the glacial climate slowly came on, the breeding range in response slowly became more and more southerly as the cold spread southwards. Species simply retreated by extermination from those glacial conditions along the line of their past emigrations or extension of range; in other words, the range shrank back upon itself. *Birds, be it remarked, do not extend their range or emigrate to increase the area of their winter quarters.* Paradoxical as it may seem, it is nevertheless a fact, that BIRDS EMIGRATE AND MIGRATE SOLELY TO BREED! Extension of range is prompted exclusively by increase of breeding population, and therefore takes place in spring, north in the Northern Hemisphere, south in the Southern Hemisphere. Hence we find birds much more sparingly dispersed in summer at their breeding grounds than in winter at their winter quarters, where they are often gregarious and exceedingly crowded or abundant in small areas, due especially to the influx of the young. Whether their winter area is large or small depends entirely upon the localness or otherwise of those conditions which were favourable to breeding occupation prevailing during the time their northern range was curtailed by glacial climates, or as the breeding range spread across the world. The Turnstone, for instance, as an example of past widespread southern breeding areas, and therefore present extensive winter range, is a circumpolar species breeding chiefly on the shores of the Arctic Ocean; yet in winter its migrations extend over all the coasts of the Southern Hemisphere, which it inhabited as a breeding species in emigrating north, and which there is much evidence to suggest it still continues occasionally to use for reproductive purposes, for it is said to have bred on Lord Howe's Island, Robben Island off the south coast of Africa, on the coast of South-west Texas, on islands in the Red Sea, on the Balearic Islands, the Canaries,

Azores, and Jamaica. On the other hand, as an example of past local southern breeding areas and therefore present limited winter range, we have the Rose-coloured Pastor, a species whose breeding range now extends from Italy to Lake Saisan in Central Asia, but whose winter quarters are entirely confined to India.

The southern limit of that ancient breeding range is now marked in a great many cases by the southern winter limits of a species. In other cases the southern limit of the Glacial (and of course Pre-Glacial) breeding range is marked by the area occupied by a sedentary southern form or representative species, whose segregation is often due to isolation from the northern form during the non-breeding season of the former. With the Inter-polar species such as the Knot, Turnstone, Curlew, Sandpiper, etc., the birds continue to go as far south towards what were once their Antarctic quarters, as they can find land on which to rest; and these movements unquestionably indicate that at some period the migrations of those birds were absolutely Inter-polar—the birds breeding at one pole and wintering at the other. With Inter-hemisphere species such as the Swallow, the birds continue to go far south into the Southern Hemisphere; and their movements also indicate that at some remote Pre-Glacial era their breeding grounds were situated low down in that hemisphere, say in South Temperate latitudes, and their winter quarters high up in the Northern Hemisphere.

At the climax of the Ice Age the breeding grounds of northern species, and the winter quarters of such species, no matter how far south they may have extended, were in the great majority of cases continuous and overlapped. Migration then would be limited, the birds that visited the northern areas in the immediate neighbourhood of the ice-sheets and snow-fields during the short summer, would draw south in winter (just as the Song Thrushes that now breed in Scotland move south in winter), but the majority of the individuals would be more or less sedentary. As the climate changed, and a return to more genial conditions commenced, the northern breeding range gradually expanded by emigration, and the migrations as gradually became longer, in many cases the breeding and wintering areas becoming discontinuous (although continuous by Migration), owing to change of climatic conditions, the North becoming more suitable for successful reproduction and summer residence, the South less so, but still adapted for winter residence.

All these facts tend to prove that Routes of Migration are not only continuous, but that they were formed whilst the species was extending its northern area by emigration, and therefore represent the expansion of breeding range which has taken place during Post-Glacial time. The various Routes of Migration to the British Islands may be divided into several very distinct classes. First we have the Routes followed by Summer Migrants—or of birds that come to our area in spring, and depart from it in autumn. These are all situated on our southern coasts. Then we have the North Sea Routes, which are chiefly traversed by the great stream of east to west migrants that visit us in autumn and leave us in spring; whilst, lastly, we have the Routes followed by species that come to our area to winter, or that pass over it on their way to more southern lands in autumn or to more northern lands in spring.

We will first consider the Routes followed by our Summer Migrants, and endeavour to trace the various points at which these species enter and depart from the British Archipelago. Unquestionably these routes are the most difficult of all to trace, owing to our sad lack of necessary information. In the first place, we must continually bear in mind the Law that forbids a southern emigration, and which renders a flight south in spring impossible. So far as I can at present ascertain, the bulk of our Summer Migrants enter the British Area between say Beachy Head and Dover. Less important points of entry are situated on all the great southern land projections from Selsey Bill to the Lizard, including the Isle of Wight, St. Alban's Head, Portland Bill, and the whole of South Devonshire (locally known as the "South Hams"). As we progress west, however, the Migration gradually assumes a weaker and weaker character, and probably is extremely slight at the Lizard. The Migration that enters by the south-west coast of Ireland, judging from the British Association Migration Reports, is weak, as we might naturally expect to be the case, owing to the wide water area between that district and continental land. We have many indications that the migration of birds into the south of England is much weaker in the west, where the sea is so much wider, than it is in the east, where the sea-passage is narrow. But little migration is reported from the Start Lighthouse, the keeper stating "that very few birds are observed at his station." For instance, many summer migrants are notably rare in the south-west of

England, or even absent altogether from that area which are commonly distributed or at least present in more easterly localities. Among such species we may mention the Redstart, which is very local west of Somerset; the Wheatear, which is only seen sparingly in spring, although further east it literally swarms all along the area of the Downs; the Whinchat, which is equally local in Devon and Cornwall, and only known on abnormal flight on the Scilly Islands, but abundant and widely distributed in more easterly localities; the Nightingale, absent altogether west of Somerset; the Lesser Whitethroat, very scarce and local in South Devon, and only an abnormal migrant to Cornwall; the Reed Warbler, absent from the entire south-west, except as an abnormal migrant; the Tree Pipit, decidedly more rare and local in the south-west than elsewhere; the Pied Flycatcher, only in limited numbers on migration (I have seen one example in four years), and might almost be classed as abnormal; the Wryneck, much more rare and local in the west than the east; the Turtle Dove, rare east of Devon; the Corncrake, much rarer and more local in the west than the east; the Kentish Plover, normally entirely absent, although one might have expected it to pass our entire southern coasts on migration; the Wood and Green Sandpipers, much less frequent in the south-west than in the eastern counties. I may here remark that the species entirely absent from the south-west of England, and obviously only entering our area further east, would have to migrate south in spring to reach that district, an extension of area contrary to Law; whilst the extreme localness of other species proves that they do not enter further east and then migrate south-west, as the much wider sea-passage is not only more fatal to the birds that cross it, but is essayed by a vastly less number of individuals. If these birds were equally common in the south-west of England as in more northern and eastern districts, we should find either a southern movement after entering our area, or as strong a migration across the wider portions of the English Channel as across the narrower portions, slightly more north and much further east—two assumptions which have no facts whatever to support them. From the above facts we may safely draw the inferences that *not a single species exclusively entering the British Islands east say of Portland Bill breeds south of Dartmoor; that all species do not breed anywhere south of their point of entrance; and that all species breeding in the extreme south-west of England enter from continental land south of them*, probably by way of Cape la Hague or the Channel Islands. There can be no doubt whatever that headlands are the great points of arrival and departure; they are the most enduring portions of the coast-line; the most important land-marks, and easily recognized. From the various southern promontories the great stream of spring migration spreads into our islands, fan-like northwards, east and west.

From the south of England we pass to a consideration of the Routes to Ireland. This island is far more isolated from continental land than England and Scotland, and the fact is reflected in the migration of birds to its area. The only possible points of egress for Summer Migrants are in the south, and separated by wide water areas from England. From a careful study of the British

Association Reports it is very palpable that there is very little migration into Ireland in the extreme south-west, infinitely much less even than that observable in the south-west of England. Records from Fastnet indicate that the Wheatear, "Swallows," and Whimbrel enter Ireland by that route, but the migration at this station is evidently trifling, especially in spring: in autumn more birds are observable, but many of them are obviously abnormal migrants too far west of their usual course south, or they should be observed in spring as well. At the next station east (Galley Head) much the same state of things prevails, the keeper very significantly remarking (*Report*, v. p. 86) that he has "never been at a station with less birds about than this one." The same remarks practically apply to all the stations on the south coast of Ireland until we reach the vicinity of St. Georges Channel, when we begin to obtain conclusive evidence of a regular and fairly strong migration in spring from the south. This water area, I need scarcely point out, is the narrowest sea-passage for all birds entering Ireland from the south, and is the point of former land connection between that island and England. It is the one principal route of Summer Migrants into Ireland; Holyhead, St. Bees, and Menai showing no spring route to Ireland whatever. The route from England either follows the extreme south-west, perhaps crossing the Bristol Channel *viâ* Lundy, but mostly higher up in the narrower portions, thence following west through South Wales to Pembroke, where the sea-passage begins; or, it may extend north-westerly from the south coast of England across the Bristol Channel or the mouth of the Severn, and thence *viâ* South Wales by the same route. That some birds follow this route to Ireland exclusively is proved by the date of their arrival in the extreme west of that island. Unfortunately we have not much data to go on, but I notice among one or two other instances that the Spotted Flycatcher does not reach the west of Ireland until the last half of May, a fortnight or more after its appearance in the south-west of England. Did it enter Ireland in the south-west it would arrive as early as in England. From recording stations in St. Georges Channel we also have many records in spring and autumn of such common summer migrants as Ring Ouzels, "Swallows," Willow Wrens, Wagtails, Cuckoos, and Corncrakes, passing north-west into Ireland at the former season, south-east on the return journey at the latter season, *viâ* England south to their winter quarters. Unfortunately we know little of the distribution of birds in South-west Ireland, *but there can be no doubt that none of the species that enter the country exclusively across St. Georges Channel breed south of lat.* 52° 10′. Species that do breed south of that latitude in Ireland enter the island by way of the coast of Cork and Waterford. As we have already seen, very similar conditions prevail in the south-west of England, the geographical position of the two areas being almost precisely alike in relation to the nearest land masses south of each.

I may here again take the opportunity of pointing out, that notwithstanding the much more favourable geographical conditions in the north-east of Ireland, not a single summer migrant is known to enter that area across the North Channel, another very convincing proof of the Law which forbids a southern extension of breeding area. It is significantly enough reported by the keeper of the South Maidens Light, in the very centre of this North Channel, off the Irish coast, that "no birds strike the lantern in April and May." This fact also implies that none of our summer migrants reach Scotland by way of Ireland.

The routes followed by these Summer Migrants very plainly indicate the physical changes which have taken place in the British Area, and also the approximate date of the first arrival of such species as emigrants in Britain. In the first place, we know that migrants continue to cross wide water areas in such cases where the line of Migration had become established before submergence had made any appreciable alteration in the physical character of such areas—hence the persistent passage of birds across the North Sea from and to our islands. In the second place, we have very ample proof that Emigration or the extension of breeding range is rarely if ever attempted across wide water areas, such increase of summer area having perhaps invariably been accomplished before submergence took place and such water divisions came into existence. Keeping in mind these two very important conditions, we are able to deduct the following interesting facts from a study of the known Routes of migratory birds in spring to our islands. First, that the birds which cross the widest water areas are the descendants of those species that were the first to colonize or extend their breeding range to our area after the Glacial Epoch. At the present day these are few both in number of species and individuals, and they are found still to continue to enter Ireland in the extreme south-west, having reached that area when the land extension southwards was much greater than it now is. They are hardy northern species, be it remarked, the Wheatear, "Swallows," and Whimbrel for instance, all birds that have extended their breeding range to the Arctic regions. The greater difficulties of such a route at the present time may possibly have reduced the number of individuals very considerably, and we can almost predict a future time when all migration by that route may cease. *If this should happen, the Wheatear will cease to breed in the extreme south-west of Ireland.* Second, that the St. Georges Channel land connection with Ireland continued to endure for some time after the land at the extreme south of Ireland had disappeared. At this time too we may be tolerably certain that much of the Bristol Channel was then dry land. These facts are still reflected in the great and almost only Migration Route to Ireland at the present day across St. Georges Channel. It marks the point of entrance into Ireland of most of its avian summer visitors. Wide water areas were already formed in the north of the Irish Sea, and submergence was rapidly spreading south down that sea—checking all emigration from Central and North Wales to Ireland, even of sedentary species, as the few last Summer Migrants succeeded in extending their breeding range to the Irish Area. Complete severance of Ireland from England must have taken place before the island received its due share of these summer visitors, the last land connections in the south across St. Georges Channel disappearing and checking all emigration in that direction. Migration, as we know, would still continue across the slowly widening sea, for once the habit of visiting Ireland was acquired, no reasonable amount of slow physical change would arrest it.

We now pass to England. There can be no question that a land connection endured across the mouth of the English Channel after the submergence which took place off the south coast of Ireland, and that which severed Ireland from England. All the evidence suggests that this submergence took place, both east and west of the British Islands, from north to south, as we shall also learn when we come to study the routes across the North Sea. Whilst the mouth of the English Channel remained dry land, a few more of our summer visitors succeeded in extending their range northwards to England—among these later arrivals we may mention the Redstart and the Tree Pipit, but the Nightingale, the Reed Warbler, and other eastern species could not have done so. The Channel Islands remained continental until the sea had encroached some distance up the English Channel. As the sea gradually encroached upon the land, and the English Channel began to appear, other birds gradually extended their range northwards across what are now the narrower portions of the Channel, and their line of migration was continued across the widening water area year by year, no single generation of birds, of course, being able to perceive the slightest change in their route. Finally, and at a much later period, when the climate must have grown considerably milder, the land mass of the Channel could not have extended much further west of the Isle of Wight; and across this narrow neck all the latest species to arrive made their final entry before the waters of the Strait of Dover mingled with those of the North Sea across the low watershed, of which the cliffs at Calais and Dover then formed the continuous highest ridge, and completely isolated the British Islands from continental land.

With such facts as these before us, surely the crossing of the Channel by migrant birds in spring and autumn can be nothing very wonderful after all. The submergence took place very slowly; the doomed land perhaps being swampy at first, then gradually studded with lagoons, and finally becoming open narrow sea, wider and wider. Throughout all this slow and gradual change the migration (and emigration) of birds went on, no single generation noticing a change, until the complete transference of land to water had been accomplished. The birds with dogged perseverance and admirable persistency stuck to their old routes—the only ones, be it remarked, of which they could possibly have had any knowledge—and continue to stick to them down to the present day. The Wheatears that cross the wide expanse of sea to the south-west of Ireland know of no easier and safer passage; the Redstarts that land in England on the South Hams of Devon know of no narrower route further north and east; whilst those fortunate individuals that have descended from the later settlers which entered our area where the Channel now is narrowest, never experience normally the greater terrors of the prolonged flight across that Channel in its wider aspects. I may also remark that there can be little doubt that the reason the stream of migration is so dominant at the Strait of Dover is purely because the land connection there was coincident with the dominant line of northern extension into our area, and that entrance at this former land connection was sufficiently south to admit of normal range extension throughout the greater part of the British Islands say north of lat. 50¾°—a fact which is proved by the strong migration east and south-east in autumn towards the Strait of Dover. The same remarks apply to the migration (and past emigration) across St. Georges Channel, only the latitude will then have to be extended north to 52° 10′. To say that birds in autumn, for instance (the same remarks apply to spring (as well) are all making for the narrowest sea-passage to the Continent is absolute nonsense. If birds were guilty of breaking the Law of their dispersal so flagrantly, why, I ask, do not the tens of thousands that cross the wide North Sea each season—between say Heligoland and Hull—pass south to Calais before they attempt to make the passage? We know, of course, that numbers do so cross at that narrow passage, but the circumstance is entirely due to the line of emigration followed by the ancestors of those individuals, and is merely a coincidence.

Palmén's elaborate system of "Fly Lines," which he postulated in his endeavour to trace the Migration Routes of birds, are myths. No special route of migration is traversed; species follow the course that their range expansion has taken in past ages; and the route can only be regarded as a *common* one (as a "Fly Line" in Palmén's meaning), in the sense that many species have extended their areas of distribution along it. It may be urged that in all countries traversed largely by migrants, there are certain routes which are much more crowded than others; that in some districts little or no migration is apparent, whilst in others it is palpable even to the most cursory observation. This is perfectly true, and is the result of the survival of the fittest, the easiest routes being occupied season by season, by the descendants of birds that extended their area along them ; the more dangerous routes being abandoned, because the perils in following them have been so much greater, and resulted in the gradual extermination of the individuals and their descendants that had followed them during their Post-Glacial emigrations. We may safely presume that in earlier ages migration was most numerous and important across certain areas (the mouth of the English Channel, and to the south and south-west of Ireland, when dry land) where it is now least apparent and least extensive, seeing that the areas to which those more ancient routes led were probably colonized much earlier and in greater abundance by emigrating or range-expanding species directly after the Ice Age, than more northern and eastern areas, say where the English Channel is now the narrowest. As the sea has encroached, the migration route has become more perilous, with the inevitable result that birds have been exterminated in their persistent efforts to follow it. At the present time the most used routes are those where the passage is attended by the minimum of danger, and where the conditions of flight are the most favourable. But this is not *choice* on the part of the individuals following these routes, but the accident of direction of Post-Glacial emigration, the survival of the fittest. We can thus see why the narrowest seas are crossed in greatest numbers by migratory birds, why the most favoured valleys, coast-lines, or mountain passes are followed, why headlands and peninsulas are more favoured points of arrival and departure than deeply indented coasts—the simple explanation being that the more difficult routes have become deserted through the extinction of the individuals following them, or continue to be followed by a few survivors, destined, nevertheless, sooner or later most probably to complete extermination. Had all routes of passage remained the same, continued unchanged so far as the dangers are concerned, Migration would have endured uniform in character between breeding grounds and winter quarters or range bases; but the various physical mutations along the route have minimized or increased the perils of the journey, and at the same time rendered that Migration uneven in character, dominant in some places slight, or even entirely absent in others. Thus the migration across the North Sea[1] is general and very uniform in character; that across the English Channel is local and most irregular.

The North Sea Routes, which are perhaps most conspicuous on the eastern coasts of the British Islands, have already been dealt with at considerable length in a previous chapter, so that little more respecting them need now be said (*conf.* pp. 130-131). We may, however, remark that they originated in precisely the same way as those Routes across the English Channel, and that the facts they now present are entirely in harmony with

[1] It is not improbable that the Migration across the North Sea may ultimately become of a much weaker character, especially as regards small Passerine birds. The area which this Migration drains is many times larger than the British Isles, and consequently the movement will continue unaffected to any appreciable extent by adverse conditions for a much longer period.

the progress of that vast submergence which brought the North Sea into connection with the English Channel across the Strait of Dover. As previously stated, this submergence undoubtedly took place in a north to south direction, if the geographical distribution of birds is of any value as an indication of past physical change. It follows then that this area, say north of the Humber, was for the most part a water area between the latter locality and Denmark, before the emigration of species eastwards commenced, yet not before a line of north-east Emigration had progressed towards Scandinavia, as is proved by the present limits of the great East to West Migration that now breaks upon the east coast of England in autumn. The bulk of this migration from the east strikes our eastern coast-line from Yorkshire southwards, just as the bulk of our northern migration in spring of summer visitors is most dominant in the vicinity of the Strait of Dover—eloquent proof, I take it, that the North Sea plains endured longer in the south than in the north. This East to West (or more correctly speaking, West to East) Migration extends right across England and Wales to Ireland; and here we may remark, that there is evidence to prove that the Irish Sea is crossed in a direct line east to west,[1] say from Holyhead southwards to the north of Pembroke, which is just what we ought to find if our previous view of the past physical changes in the British Area is a correct one. The birds that follow this east to west

[1] Here, for instance, is the report of the keeper of the Kish Bank Light Vessel, stationed off Dublin Bay: "September and October are the chief months for the migration of birds from the Welsh coast" (*Report*, iv. p. 76).

route in autumn are all hardy species which must have been among the first to enter Britain after the Ice Age had passed away, and when it was practically a compact continental area. No Irish Sea then existed; Ireland was then probably better suited to the requirements of all Passerine species than any other part of Britain, due entirely to its proximity to the open sea and warm ocean currents, and from there emigration eastwards undoubtedly commenced, as is proved by the migration to that area of so many of these east to west species. The Starling, the Sky Lark, and the Short-eared Owl are especially good instances of this line of past Emigration, as they migrate in such vast numbers, palpable to every observer, right across England to Ireland, where they are known to occur in enormous quantities in autumn, starting on their return journey east in spring. I may also remark, again, that the Short-eared Owl is not known even to breed in Ireland now, although common enough there in winter. It may also be stated that none of this east to west migration is perceptible in the south-west of England, although there is a considerable amount of the migration from the north-east observable even down to Land's End (*conf.* table, p. 132). Here, then, is eloquent testimony of the impassable nature of a water area to Emigration, and of the futility of such an obstacle to arrest Migration established previous to the submergence! We have seen how Emigration has been effectually checked by the comparatively narrow St. Georges Channel, and now see how the wide Irish and North Seas—formed after successful emigration—are impotent to arrest the Migration of species whose range was gradually extended over their area when dry land. The submergence which caused the incroach of the Irish Sea has had no more visible effect upon this East to West Migration than the subsidence which drowned the North Sea plains has had on its progress there. The habit of crossing those areas has been slowly acquired by an extension of breeding range, and still continues to be followed with a persistency as astonishing as that which characterizes the migrations of the lemming, whose impulse to migrate over the now submerged or effaced lines of its former emigrations is so dominant that nothing but death itself can eradicate it! In conclusion, we may remark that evidence of the north-east to south-west migration to Ireland is evident at many stations in the North Channel and in the north of the Irish Sea, one of the most interesting instances (because so easily traced by numbers) being that of the Goldcrest; a species which, we have already shown, extended its breeding range north-east across the British Area to Scandinavia. These north-eastern species are specified in a preceding chapter (*conf.* p. 132).

We now pass to a consideration of those Routes followed by species that come to our islands to winter, or that pass over them on their way to more northerly or southerly areas. Broadly speaking, these Routes extend very impartially over the United Kingdom, as we may very naturally expect to be the case. All these species are northern in their summer dispersal—hardy or Polar species that were absolutely the first to extend their emigrations over the British Area after the climate of the Ice Age had sufficiently moderated to permit of successful avian colonization northwards. They are species that at one time bred in our area, and whose range extended northwards with the changing climate across it, until Britain was entirely (or nearly) deserted in summer, as the breeding range more or less completely passed on to the north. I say "nearly" in some cases, because a few individuals of such species for instance as the Wigeon, the Greenshank, etc. (*conf.* table, p. 151), still continue to breed in our islands, but more dominantly to winter in them or to pass them as coasting migrants. None of the birds whose breeding and winter range overlap in the British Area are Inter-polar or Inter-hemisphere. Some of these birds, as for instance the Little Stint, the Sanderling, and the Ringed Plover, were never resident in our area (no more than the Swallow is), although their breeding range once unquestionably extended over it, and as far to the south of it as the winter quarters once extended north from south Polar latitudes during the time Antarctica was occupied. They are dominant Inter-polar species that must always have passed far to the south or south-east to winter, and which may be regarded as our earliest Post-Glacial summer migrants, whose breeding range now only begins far to the north of us.

During this period the British Area was practically unbroken and compact, not only far south towards the Bay of Biscay, but far north in the direction of Iceland, Greenland, and Scandinavia. The vast submergence that has taken place has broken much of the continuity of these Routes of Migration, but they are still followed, as we have found to be the universal rule. The North Sea between Scandinavia and Scotland is still swept by these migrant hosts each season: the wider waters are still crossed between Greenland, Iceland, and the Faroes, and between the south of Ireland and the Lizard, as they were in remote ages traversed when dry land, or nearly continental dry land, replaced them. Most of these birds, it will be remarked, are aquatic, able to make long flights across the sea, powerful of wing; and it is possibly due to these facts that they have been able to conform to the Law of Migration so successfully and so long, notwithstanding the enormous change which has taken place along their recognized routes—now only rendered visible in some cases by an isolated island here and there, which may serve as a welcome landmark or a possible place of rest. We thus see again that the wide water areas are no obstacle to species that acquired the habit of crossing them when they were dry land, although they are barriers to emigration and range extension which no species attempts to pass.

These Routes unquestionably are most followed along our eastern coasts, due not only to the greater land mass of Scandinavia lying nearest to that area, but to the fact of the Shetlands assisting in deflecting a great deal of it east. Most of the species that follow them are coast birds, and they chiefly keep to the coast-lines on their way to their northern or southern destinations. We have one important Route from Scandinavia, not only by way of the Shetlands and Orkneys, but right across the North Sea in a south-westerly direction to the Scotch coast. The former portion of this Route divides at the Orkneys, one continuing down the east coasts of Scotland and England, the other following the west coast of Scotland to Ireland and the west coast of England. Another Route, followed, however, by few land birds, extends from Greenland, *viâ* Iceland, the Faroes, St. Kilda, and the west coast of Scotland, to Ireland. I have already dwelt at some length upon the emigrations of species that followed many of these Routes, so that it is unnecessary to enter into greater details here.

Before dismissing the subject of Routes it will now be advisable to deal a little more fully with their inland continuations. These, I need scarcely remark, are extremely difficult to trace, owing to the sad lack of an all-necessary series of long-continued observations. There can be little doubt, especially in the case of locally distributed species, that these Internal Routes are complicated, although they will be found to conform exactly to those laws which govern the migration of birds elsewhere. We have already seen that birds follow most closely in their present migrations the past line of Emigration—however complicated and tortuous that may be. Could we only define the limits of emigration, we should find them to agree precisely with the present Routes of Migration. We may safely put down as an invariable rule that the Internal Routes of migrants trend in the present direction of suitable breeding conditions, or over such districts that were once suitable areas for reproduction. River valleys, as long as they trend northerly, are certain routes for migrants ; for there can be no doubt that from earliest times they were favourable localities for range extension. Mountain chains will just as surely indicate the routes followed by other birds whose conditions of successful reproduction are only to be found in upland districts ; whilst lake systems, swamps, heaths, woodlands, or cultivated districts will all mark the Route of Migration followed by the several species that reside only in such localities. The migration routes of the Dotterel, for instance, will, broadly speaking, follow the mountain uplands; and this may explain why so little of this species is seen whilst it is on passage. It is also interesting to remark that this bird is known to pass on migration the districts where it formerly bred. The Ring Ouzel, the Wheatear, and the Merlin will also follow a mountainous route, as the direction along which their past emigrations were carried. The routes of such species as Sandpipers and Ducks will follow rivers and lakes; those of the Rails and Snipes will follow the northern trend of the swamps; those of the Stone Curlew and the Nightjar the heath systems and the commons; whilst those of such species as the Hobby and the Honey Buzzard would be confined to the northern extension of woods. In many cases these Routes have lapsed as Breeding Routes, either through modification of climate, as the northern extension of breeding range spread over our area, or the disappearance of suitable breeding grounds, as such have been "improved" away. Such Routes will continue, however, to be followed by the descendants of birds that increased their range along them. In the utter absence of detailed information, I shall refrain from more closely entering into the subject of Internal Routes; but I may refer the reader, in addition to the above information, to what I have already said in the *Migration of Birds* (pp. 242-245).

CHAPTER VIII.

CONDITIONS OF FLIGHT.

Routes of Migration, how followed by Birds—Paley's Definition of Instinct—Impulse of Migration—Restlessness of Captive Birds—Certain Routes followed by Certain Individuals—How a Route of Migration has been Learnt—Mysterious "Sense of Direction" a Myth—Altitude of Migration Flight—Advantages of a Lofty Course—The Order of Migration—A few Old Birds Migrate as Early as the Young—The Daily Time of Migration—Amount of Sociability amongst Birds on Passage—The Perils of Migration.

IN the preceding chapter we endeavoured to ascertain how a Route of Migration has been formed; it now becomes necessary to inquire how birds continue to follow those routes so unerringly, how they manage to traverse them for such long distances apparently with so few mistakes. If we define Instinct as Paley did, and describe it as "a propensity prior to experience and independent of instruction," which I think is about as good a definition of the power as we can ever hope to possess, then most assuredly Instinct can never control the performance of avian season-flight. The impulse of migration may be, and probably is, a very deeply rooted one, an hereditary impulse even. Captive birds in which the habit of migration is dominant, have often been observed to become exceedingly restless as the usual time for their departure approaches, and there is much evidence to suggest that such restlessness is invariably increased or further excited by the sight of companions *en route*, or by the cries they utter when on flight. We remark the display of a similar state of restlessness among birds at liberty upon the eve of their departure, the gatherings of some birds, the unusual activity of others. Whether a falling temperature or a failing food-supply assist in intensifying this impulse we do not at present know, but there can be little doubt that the departure is taken when the impulse becomes too intense to be longer resisted, whatever be its initiating cause. Once, however, a bird begins its migration all Instinct as a guiding medium ceases; memory and knowledge of locality, in fact experience, assist it to perform that long journey. Migratory birds follow routes in every case that indicate the line of their extension of range or past emigration. These routes have been slowly formed and are continuous, either from the area where the species now winters, and where in past ages it formerly bred, or by the absolute overlapping of the summer and winter range. The individuals that follow one route by no strange chance ever follow another; even though their own may be fraught with perils and difficulties unknown to the other, it is still retained; and even though one route may be much longer and more tortuous than the other, it still continues to be followed. If this were not so, why should some individuals of the Wheatear, for instance, elect to take a wide ocean passage to the south-west of Ireland, whilst other individuals cross by the Strait of Dover? Why should the Redstarts breed-ing in or passing over Devonshire, cross the English

Channel where it is four times as wide as where individuals breeding in or passing over more eastern counties cross? There can be but one answer to such a question, and that is, that these individual birds follow a route along which their ancestors increased the range extension northwards, that they have no knowledge of any other route, and are not endowed with any instinctive faculty that will help them to migrate by any easier way. If the route be exceptionally dangerous, or the difficulties in following it increase, it will still continue to be followed, until every bird that follows it is exterminated; it can never be changed (*conf.* pp. 223-226).

The Route has been slowly learnt by the species as the summer area increased. When the extension of breeding range commenced the extent could probably have been measured by yards rather than miles; by a short flight to a more distant lake or swamp; a visit for nesting purposes to some outlying grove or forest; or a trip north along a coast for a little way to a suitable stretch of sand or shingle, or range of cliffs—prompted chiefly by overcrowded quarters to the south. As the species multiplied the range increased. Each year the journey became longer, more distant from the base or centre of dispersal; so slowly that perhaps not more than a mile or so might have been added to the northern area of a species in a century, as favourable conditions to extension were gradually presented. The flight back in autumn was therefore at the beginning a very small one, and never perhaps included any sea-passage whatever. But our Summer Migrants now, we know, cross the Channel every year; they do so because the line of their past emigration extended across that area when it was dry land; their migration route gradually assumed its present aspects as submergence progressed, no single generation of birds having experienced any very sudden change, the sea-passage becoming imperceptibly wider and wider as the land vanished, until the present state of things was reached.

That migration routes are traversed by experience, and not by inherited impulse, is further proved by the variability of the habit, some individuals going longer distances than others, some remaining stationary altogether. Again, if birds are endowed with that mysterious "sense of direction," which popular opinion so readily ascribes to them, how can that sense explain the endless routes, the tortuous journeys, the migrations this way and that to common range bases or centres of dispersal? How utterly at fault it must be in those species that breed in the far north-east and that winter in the remote south-west; how impotent in species that breed in Pomerania and journey south-east all the way to India to winter, when just as suitable localities are available directly south and not a quarter of the distance! Is it not more rational to presume that these migrant birds are following the route of their ancient range extension, a route with which they must be thoroughly familiar, the result of it may be thousands of years of experience? The very fact that migratory birds, generally speaking, keep so closely to their normal areas of distribution is a most convincing proof that they do not wander from their routes of passage. For this reason I consider it most absurd to say that a wave of migration has been deflected this way or that normally. If it were so, birds would be drifted into country of which they have had no experience, of which they can therefore have no knowledge, and would be lost to all direction or locality. How rarely do we have any proof of this. It is true migrants are repeatedly driven out of their usual course by storms and fogs, and this abnormal deflection from their proper route is fatal to vast numbers of birds every year. Were birds endowed with this mysterious inherited sense of direction, most of the wonderful scenes at our lighthouses in spring and autumn would never be witnessed at all. But the perils of the journey are great and constant; birds blunder to an almost incredible extent; lose their way or perish every year in numbers that can only be described as astounding!

Another very important condition of Migration which is too often and too persistently ignored by the majority of people is the altitude at which it is undertaken. I have already dwelt at considerable length upon this subject in the *Migration of Birds* (*conf.* pp. 77-84), to which I would again refer the reader. The advantages of such lofty flight are very obvious. Let a person stand on some moderately lofty hill, say 400 feet or so above the level of the sea, and let him remark the vast area of country over which his vision can extend. The more he increases his altitude, the wider will become the view—hills, valleys, and coast-lines spreading out before him in one uninterrupted expanse. Birds flying along a lofty course will readily recognize the various landmarks that they and their ancestors have been in the habit of passing for ages. A headland, a river valley, a wide reach of sand or shingle, for instance, may mark the spot where the sea-passage begins; a lofty down, a similar headland, or a wide expanse of heath or forest may just as surely indicate the point where that sea-passage terminates. Inland Routes are followed in much the same manner, old familiar points being recognized and passed with amazing precision every year. I can recall many instances of birds passing certain moorlands, certain sheets of water, certain woods and heaths on migration with a regularity that must surely indicate the appearance every season of the same individuals or their direct descendants.

A few words now on the Order of Migration. There is a very generally prevalent idea that the young birds are the first to migrate in autumn, and this has been repeatedly brought forward as a most convincing proof that birds are born with an instinctive knowledge of the route they must traverse to their winter quarters. That this order of migration prevails amongst many species cannot be doubted. The young unquestionably appear in many localities it may be weeks in advance of the general migration of adult birds. But let it be remarked that a few old birds invariably precede as well as accompany these first flights of the young—old individuals with a full knowledge of the road that have acted as guides to the inexperienced. Many young birds, however, go astray on their autumn journey south, and it is remarkable that the bulk of abnormal migrants at that season is composed of young birds that have lost their way. As I previously pointed out in the *Migration of Birds*, the individuals that are the first to migrate in autumn are birds that have been prevented from breeding or that have lost their broods. Such individuals have no parental instinct to restrain them from starting early, and they frequently begin to move south before their moult is completed. These may be regarded as the pioneers, and with their departure the flight south of the young commences. These young often set off as soon as they can fly, and individuals of certain high Arctic species have been observed on the British coasts even with the down of their nestling plumage still adhering to them. Soon after the departure of the bulk of the young the adult males begin to leave their summer quarters, the females following a little later, their moult being delayed somewhat by maternal duties. In the rear of the migration come the laggards—individuals delayed by accidents to their flight feathers, or other casualties. The order of return in spring is to some extent reversed. As usual, the adult males initiate the migration; then follow the females; the young of the preceding season follow, and lastly the maimed and weakly individuals. These and the young (or birds of the last season) frequently pass the summer some distance south of the actual breeding grounds, or actually remain in some cases in the winter quarters.

The daily time of migration also varies considerably. Some species migrate exclusively by day; others just as regularly by night; some by night as well as by day. The punctuality of their arrival and departure is also profoundly interesting; and it will almost invariably be found to be the rule that the birds that arrive earliest in spring are the latest to depart in autumn; the latest to arrive in spring being the first to take their departure. I should be disposed to class the former of this class of migrants as much earlier emigrants to the British Area than the latter.

The varying amount of sociability in birds whilst on migration, their gregarious or solitary tendencies, have all been described at length in the previous volume, and as the facts apply not only to British birds, but to all species, we need not repeat them. We have also dwelt at some length on the duration of the passage and its varying phases of intensity. The migration is usually characterized by the advent or departure of a few individuals; then the flight becomes more strongly marked up to its greatest phase of intensity, which may or may not be marked by one or two exceptionally strong movements, after which the passage of the species for the season as gradually dies away as it commenced. The various stages of the journey, and the normal velocity of flight during passage, have also been described. I have nothing further to add at present to what I have already said in the above-mentioned work concerning the effects of Wind and Temperature on the migration of birds. The remarks just as aptly apply to British birds as to other species. The moulting of birds before they migrate, and the structure of the wings of migratory birds, have also been dwelt upon.

Of the perils of migration but little more may be said. I have already devoted a chapter of my previous work to the subject. Many more instances of fatalities to migrants might be given, but they would all partake of the same general character: sufficient has been furnished to illustrate very vividly the perils that beset birds whilst on passage. The mortality connected with migration can scarcely be realized. The vast numbers of birds that perish during passage is past belief. And this not only applies to the British Area but to every other part of the world. There can be little doubt that the passage across seas is attended by the greatest amount of mortality, and even under the most favourable circumstances the death-rate must be a high one. Of the hordes of young birds that are seen on migration in autumn but a small percentage survive the perils of the journey, and a still smaller percentage return in the following spring. If the season of passage chances to be exceptionally unfavourable—full of storms and fogs —the death-rate will go higher than its normal average, and perceptibly reduce the numbers of a species perhaps for years. An anonymous reviewer of the *Migration of Birds* says that I greatly over-estimate the perils of migration. Well, after several more years' close attention to the subject I will state that in my previous work I under-estimated the mortality, and that the more I study migration the more I am convinced of its fatal consequences.

CHAPTER IX.

THE SPRING ASPECTS OF MIGRATION IN THE BRITISH AREA.

Commencement of Spring Migration in the British Islands—Departure of Eastern Migrants—Departure of North-eastern Migrants—Abnormal Lines of Migration from the British Area—Birds Migrating Too Early—Arrival of First Spring Migrants from the South—Departure of Winter Visitors to the British Islands—Coasting Migration in Spring—Migration of various Northern Birds—Arrival of Summer Migrants in the British Islands—The Growing Intensity of Spring Migration—Months of Passage of Various Species—Gradual Advance Northwards of Migrants—Duration of Spring Migration—Vertical Migration in Spring—The Various Species performing it—Order of Migration—Table indicating the Spring Migration of Birds across the British Islands.

I INTEND to devote the present chapter to a brief review of the most salient features of the Spring Migration of Birds as they are presented in the British Archipelago. To the student of British birds, nothing perhaps is more interesting after a long dreary winter than to note the first signs of a coming change of season in the movements of migratory birds. As might naturally be expected, the first species to display any migratory tendencies in spring are the hardiest—those birds that winter in our islands, yet breed in various continental areas east and north of them. This migration in spring is much less marked than in autumn, and for three reasons. First, it is always more difficult to note the departure of a migratory species than to observe its arrival; second, the number of individuals that depart in spring is very much less than the number that arrived in the previous autumn, owing to the casualties of the intervening winter; and third, the majority of the species that take part in the movement are birds that breed commonly in our area, and it is difficult if not impossible to distinguish between the individuals that only winter here, and those that spend the summer with us.

This migratory movement, in a fairly normal spring, may be said to commence as early as February, and the first migrants to go are those that breed in continental areas east of the British Islands. Amongst these species we may include the Missel Thrush and the Song Thrush, the Hedge Accentor, various species of Titmice, the Wren, the Linnet, and some other Finches, several species of Bunting, the Jay, the Rook, and the Carrion Crow. In former years, when the Great Bustard and the Bittern were common in our islands, these species also would have been amongst the first migrants to move towards the Continent. The indication of their migration is now very slight. For a complete list of the species partaking in this early spring migration the reader is referred to the table on p. 132. The next species to move out of our islands are those whose lines of Migration trend north-east. It is impossible to separate the individuals, but many birds of these species migrate both east and north-east; those travelling in the latter direction are the last to go. Among these we may include the Blackbird, the Robin, and the Goldcrest, the Greenfinch, the Chaffinch, and the Starling, the Jackdaw and the Hooded Crow, the Short-eared Owl, the Ring Dove, and the Lapwing. Years ago, before the Crane had been well-nigh exterminated in the British Islands, it would have left its winter quarters in our area with this second migratory movement. We have abundant testimony from the various lighthouses and light-vessels stationed on our eastern coast, or in the North Sea, that the migration of all these species is undoubtedly in progress very early in the spring, continuing more or less strongly for just upon a couple of months. Sometimes birds of these species are noticed passing away from the British Area in great numbers for days together, but the migration is only in very exceptional instances so strongly marked as the return movement in autumn. It has been stated (conf. *Migration Reports*) that there is apparently a *south-eastern* departure from our shores in spring, but this, if true, is utterly abnormal, and due entirely to local conditions. I for one should require more accurate and positive proof of this movement than that furnished in support of it by the entirely unskilled observers that have reported the anomalous migration. It is impossible to say accurately in what direction a body of migrants may be going on a dark night, when the birds are bewildered by adverse weather, and flying dazed and lost round the lanterns of a lighthouse. We can only accept the reports as useful when they are made under fairly normal circumstances, and obviously harmonize with known Laws of migration and geographical dispersal. Again, a strong migration from the east and north-east to the British Area in *spring* has been reported and described by Mr. Cordeaux as a "very anomalous movement." If it were perfectly normal it certainly would deserve such a description. There can be no doubt whatever that the individuals coming from the north-east and east (most probably the former) into England, in the very teeth of the advancing normal migration from Britain to the Continent, are birds that have migrated too early, or that a sudden late snap of winter in their summer quarters, or on the routes to them, has turned back. We experience precisely the same phenomenon in the Arctic regions of birds being too eager to reach their breeding grounds, and having to return, it may be, many miles, owing to a temporary recurrence of winter weather. It would be just as manifestly absurd to say that there was a migration in the Arctic regions *south* as well as *north* in spring, as it is to describe this east to west or north-east to south-west flight at the same season on and off our eastern coasts (conf. *Migration Report*, v. pp. 60, 61). It is suggestive enough that no such "anomaly" occurs in autumn!

During the earlier part of the time this spring migration is in progress of birds leaving us for continental areas, another, if perhaps less perceptible movement of some of the same species is taking place into our islands from more southerly areas: individuals that breed with us and still continue to winter further south. This movement is not a very ancient one. It is made by the descendants of individuals of these species that were among the last to extend their summer area to the British Islands. It was initiated after Ireland was separated from England, and after much of the Channel land had been submerged. Hence no trace of that movement is visible in extreme western and south-western districts.

Very soon after this early spring migration begins the birds that are only known in the British Islands as winter visitors commence to leave and to pass northerly, either by an easterly trend to the Continent, or by a westerly trend to the Faroes, Iceland, and Greenland. The Red-wing and the Fieldfare begin to leave our islands towards the end of February, the migration of both species becoming more marked in March, reaching its fullest strength in April, and ceasing in May. The Brambling passes north again in March and April, as also do the Siskins that winter exclusively in our area. The Snow Buntings begin to leave us even at the end of January, the migration being more regular in February, attaining its fullest strength in March, and dying down completely in April. Such species as the Golden Plover, the Lapwing, the Curlew, the Redshank, the Woodcock, the Common and Jack Snipes, individuals of which winter in our area, begin to pass northwards from it in February, the migration becoming strongest in March and April, and dying away again in May, or in some few cases is prolonged even into June—the latter being composed chiefly if not entirely of coasting migrants, individuals that only cross our area on passage. In February and March great numbers of Ducks and Geese commence their northern migration from our islands, the flight continuing through April, and in the case of the most Arctic species into May and even June, although the latter migration probably consists entirely of coasting migrants—individuals that merely pass us from winter quarters in lower latitudes.

This coasting migration begins apparently as soon as the migration which is exclusively of a departure character, but is weak until March, and perhaps attains its greatest strength in April and May. It is impossible to distinguish with certainty between coasting migrants and migrants that have passed the winter with us, or in the case of species that breed in our area; but with those species—high boreal forms, as a rule—that neither winter nor breed with us, there is no difficulty, and these birds generally coast us late in spring. The Whimbrel, for instance, passes us very regularly towards the end of April, but the migration of this species is most marked in May, dying down in June. The Ringed Plover begins to pass our coasts in March, very sparingly, the great flights crossing us in April, and lesser numbers again in May. The Sanderling passes us pretty regularly throughout April and May, a few in June; the Skuas of various species coast us chiefly in April. The Little Stint passes our coasts sparingly in April, but in greater numbers (although even then by no means dominantly) during May; as also does the Knot, which perceptibly prolongs its passage into June. The Curlew Sandpiper, never very common with us, coasts us in April, May, and June.

Owing entirely to the fact that all or nearly all this early spring migration is entirely among species that breed or winter in the British Islands, no season-flight is very apparent until the species that visit us exclusively in spring to breed in our area begin to make their appearance. While all this other migration which we have been describing is actually in progress—long before it ceases—with the advent of March in fact, the northern migration of our summer birds begins. Amongst the very first of these spring migrants we must class the Woodcock, which begins to arrive in our area in February, but the migration is strongest in March and April; another early bird to appear is the Pied Wagtail, that is to say, the individuals that breed with us and winter further south. The Wheatear and the Chiffchaff, however, are the two best known species that indicate the beginning of the exclusively spring migration. Neither of these birds winters in the British Islands normally, but they reach us in March. At first the migration is a weak one, but as April advances the flight becomes stronger and stronger, and wanes gradually again into May. But little migration amongst our typical summer birds, however, is apparent until April. Almost the only species that are seen to arrive in our islands in March are perhaps the two just mentioned, together with the Kestrel, the Ring Ouzel, the Willow Wren, and the Yellow Wagtail. More rarely a very slight migration of the Wryneck, the Stone Curlew, and the Garganey may be observed during March; but with none of these species does the flight become dominant until April.

Scarcely without exception the migration of every species that visits the British Islands in summer begins to be apparent some time during this latter month. The most noteworthy exceptions are the Osprey, the Spotted Flycatcher, and the Marsh Warbler; the two latter species, as we have already seen, migrate far to the east and south of our area. With the advent of April, then, migration from the south begins to assume a gradually increasing strength, which continues to swell in volume as the month progresses, and with most species to be continued well into May. Daily the numbers of Chiff-chaffs, Willow Wrens, Whinchats, Redstarts, Wheatears, and Warblers increase. It is chiefly owing to the observations made at the points where migrants enter our area that we are able to compute the duration of the flight of these species. From these records, extending as many of them do over a long series of years, we are able to judge pretty accurately not only the duration of the migration of each species, but the time it begins and ceases. We are thus able to state that the migration of all birds that begins in April is continued into May, and much more exceptionally into June. With such species, however, as the Honey Buzzard, the Garden Warbler, the Lesser Whitethroat, the Wood Wren, the Cuckoo, the Turtle Dove, the Quail, the Red-necked Phalarope, the Common Sandpiper, and all the Terns, the migration is much stronger in May than in April. Probably all the birds that breed in our islands have reached every part of them towards the end of May. The migration, however, of some of the common summer migrants continues into June. We have ample evidence that such species as the Wheatear, the Swallow, the House Martin, the Cuckoo, the Dotterel, the Whimbrel, the Ruff, and the Arctic Tern continue to enter our area in June (the Swallow especially), but there can be no doubt whatever that these late individuals are coasting across the British Islands on their way to more northern breeding grounds, and in most cases pass by other individuals of the same species already engaged in the duties of reproduction. It is noteworthy that all these late coasting individuals not only breed in the high north, but have extended their emigrations across our area thereto, as has been already shown (*conf.* table, p. 116).

The spring migration of birds to the British Islands is just as gradual as in other localities situated in temperate latitudes. Birds are almost invariably seen first in the more southern districts; in cases where this is not so they must have been overlooked. Nearly if not quitė two months elapse before the remote northern and western areas are reached. Thus most of the summer migrants to Ireland reach that country a week or even a fortnight later than they appear in the south of England. In the north of England most migrants are at least a week later to arrive than they are in the south, and this is much more marked in some species than others. The Chiffchaff, for instance, must migrate very rapidly across our area, as it is generally seen as soon in the north of England as in the south, at most but a day or so intervenes. On the other hand, the Redstart, the Ring Ouzel, and the Corn Crake must travel slower, for often ten days or a fortnight will mark the difference in the date of their arrival. A still greater difference in the date of arrival is almost invariably remarked in Scotland. As a general rule the state of the season governs the progress of the migration. If the weather be mild and open birds reach their northern destinations quicker than if the season be a cold, stormy, or backward one.

The Spring Migration of Birds to the British Islands extends over a period of quite four months. It begins in February and certainly lasts until the end of May or the first week in June. In some cases the migration of a species extends over the entire period, as for instance the Wheatear; in other cases the flight does not cover more than three months, and even then in a great many instances the migration is weak at the beginning and at the close of that period; whilst in other cases the migration is confined to two months, especially in such species as the typical Warblers, the Wagtails, and the Pipits. Other species limit their spring migration to a month. Among these we may include the Marsh Warbler, the Red-backed Shrike, the Hoopoe, the Red-necked Phalarope, and the Greenshank. As a general rule we might say that the most widely-dispersed species cover the longest period in their migrations, the most local birds the shortest period, the length of flight-time being in the same ratio as the dispersal of the species.

A few words concerning the Vertical Migration in spring now become necessary. This migration is just as characteristic and as marked a feature of the spring in mountain districts as that taking place in a latitudinal direction. The duration of the period also is about the same, and lasts for about four months, although it is not perhaps so marked at the beginning or at the close. The movement is more contracted, more rapid. Amongst the earliest of these vertical migrants we may instance the Stonechat, the Linnet, the Gray Wagtail, the Wood Lark, the Sky Lark, and the Lapwing. These species begin to move up the hills in February, but the migration is much more strongly marked in March, and dies completely away during April. The next birds to move are such species as the Merlin, the Linnet, the Twite, the Lesser Redpole, the Pied Wagtail, the Meadow Pipit, the Golden Plover, the Curlew, and the Dunlin. beginning in March and continuing into April. With these species the movement is more contracted, in some the bulk of the flight taking place in March, in others in April. Some few species, such as the Linnet, the Twite, the Lesser Redpole, the Golden Plover, and the Curlew, prolong their migration into May, but these are individuals that breed at the highest elevations and in the most northerly localities. In our islands all vertical migration ceases in May, but in higher latitudes it is prolonged even into June.

The following table will indicate the duration of the principal Migration across the British Archipelago in Spring.

SPECIES.	To N.; N.E.; E.					From South to North.					From South to North.				
	DEPARTURES.					ARRIVALS.					COASTING.				
	FEB.	MARCH.	APRIL.	MAY.	JUNE.	FEB.	MARCH.	APRIL.	MAY.	JUNE.	FEB.	MARCH.	APRIL.	MAY.	JUNE.
Missel Thrush ...		× ×	× ×	...	...	×	× ×	× ×							
Song Thrush		× ×	× ×	×	...	×	× × ×	× ×							
Redwing		×	× × ×	×	...	...	...	...		...	...	× ×	× × ×	×	
Fieldfare		× ×	× × ×	×	...	...	...	...		...	...	× ×	× × ×	×	
Blackbird		× ×	× ×	×											
Ring Ouzel		...	...	...	...		× ×	× × ×	×	...			× × ×	×	
Black Redstart		× ×													
Common Redstart ...		...	...	...	...	...		× × ×	× ×						
Whinchat		...	...	...	...	...		× × ×	× ×	...			× × ×	×	
Stonechat		×	× ×	× ×											
Wheatear		...	...	...	...	×	× ×	× × ×	×	×	...		× × ×	×	×
Robin		× ×	×	×											
Nightingale		...	...	...	...			× ×							
Hedge Accentor ...		× × ×	× ×												
Whitethroat		...	...	...	...	...	...	× ×	× ×						
Lesser Whitethroat ...		...	...	...	...	...	...	× ×	× ×						
Blackcap		...	...	...	...	...	...	× × ×	× ×						
Garden Warbler ...		...	...	...	...	...	...	× ×	× ×						
Chiffchaff				...	...	...	× ×	× ×	× ×						
Willow Wren ...				...	...	...		× × ×	× ×						
Wood Wren				...	...	...		× ×	× ×						
Reed Warbler ...				...	...	...		× × ×	× ×						
Marsh Warbler				...	...	...			× × ×						
Sedge Warbler ...				...	...	...		× × ×	× ×						

Grasshopper Warbler	...	...	...	...	...	...		× × ×	× ×						
Goldcrest	...	× × ×	× ×	×	...	...		...	...	...	...	× × ×	× ×	×	
Long-tailed Titmouse	...	× ×	× ×	×											
Great Titmouse ...	...	× ×	× ×	×											
Coal Titmouse	...	× ×	× ×	×											
Blue Titmouse	...	× ×	× ×	×											
Wren	...	× ×	× ×	× ×											
Pied Wagtail	...	× × ×	× ×	× ×	...		× ×	× × ×	× ×	...			× ×	× ×	
White Wagtail	...	...	...	...	...	...	...	×	×	...			×	×	
Yellow Wagtail ...	...	...	...	...	...	...	× ×	× × ×	×						
Gray Wagtail	...	...	...	...	...	...	× ×	× × ×	× ×						
Tree Pipit	...	...	...	...	...	...	...	× × ×	×						
Golden Oriole	...	...	...	...	...	...	...	× ×	×						
Great Gray Shrike ...	...	× ×	× ×	×	...	...	...	...	...	...		× ×	×		
Red-backed Shrike ...	...	...	...	...	...	...	...		× × ×						
Spotted Flycatcher ...	...	...	...	...	...	...	...	...	× × ×						
Pied Flycatcher ...	...	...	...	...	...	...	...	× ×	× ×						
Swallow	...	...	...	...	...	...	...	× × ×	× ×	...		...	× × ×	× ×	× ×
Martin	...	...	...	...	...	...	...	× ×	× ×	...		...	× ×	× ×	×
Sand Martin	...	...	...	...	...	...	...	× ×	× × ×						
Waxwing	×														
Goldfinch	...	×	× ×	×											
Greenfinch ...	×	× × ×	×	×	...										
Linnet	×	× × ×	× × ×	× ×						...		× × ×			
Chaffinch		× × ×	× × ×	×	...					...		× × ×	×	×	
Brambling		× ×	× ×	×	×					...		× ×	×	×	
Tree Sparrow ...		× ×	× ×	×	×										
Yellow Bunting	...	× ×	× ×	×	...	×	×	×							
Corn Bunting ...	...	× ×	× ×	×	...	×	× ×	× ×							
Reed Bunting ...	...	× ×	×		...	×	× ×	×							
Snow Bunting ...	×	× ×	× × ×		...	...	...						× × ×	×	
Sky Lark	× ×	× ×	× ×		...	× ×	× × ×					× × ×	× ×	×	

Species.	To N.; N.E.; E. Departures. Feb.	March.	April.	May.	June.	From South to North. Arrivals. Feb.	March.	April.	May.	June.	From South to North. Coasting. Feb.	March.	April.	May.	June.
Shore Lark	× ×	×	...		...	...	...	...	...	...	× ×	×	×	×	
Starling	×		× × ×		...	×	× × ×	...	...	...	×	× × ×	× × ×	×	
Jackdaw	×	× × ×	×			...	...	...	...	...	...	× × ×	× ×		
Carrion Crow ...		× × ×	× ×		...	...	× ×	× × ×							
Hooded Crow		× × ×			...	...	...	...	...		...	× × ×	× ×		×
Rook		× × ×	× ×		...	×	× × ×								
Jay		× × ×	× ×												
Swift		...	...			...	...	× × ×	× ×						
Nightjar		...	...		...	...	...	×	× × ×						
Wryneck		...	...		...	...	×	× × ×							
Hoopoe		...	...		...		...	× × ×							
Cuckoo		...	...		...		...	× ×	× × ×	×	...		×	× × ×	×
Kingfisher		×	×												
Short-eared Owl ...	...	× × ×	× ×		...			...		...		× × ×	× ×	×	
Snowy Owl	×														
Sparrow Hawk ...	×	× ×	× ×												
Rough-legged Buzzard	...	...	×					...	...	...			×	×	
Hen Harrier	...	...	...		...			× ×	×	...			× ×	×	
Montagu's Harrier ...	...	...	...		...			× ×	×						
Hobby	...		...		...			× ×	× ×						
Merlin	...		...		...			× ×	×	...			× ×	×	
Honey Buzzard ...	...		...		...				× ×						
Osprey	...		...		...										
Bittern															

White-fronted Goose	...	×	× ×	×	...		...		× × × ×
Bean Goose	...	× ×	× ×						
Pink-footed Goose ...	...	× ×	× ×						
Bernacle Goose ...	×	× × ×	× ×	×					
Brent Goose	×	× × ×	×	×			...	×	× × × ×
Hooper Swan	× ×	×	×	×	...		...	×	× ×
Bewick's Swan	× ×	×	×	×	...		...	×	× ×
Mute Swan	×	×							
Mallard	×	× ×	× ×	×	...		...	×	× × × ×
Teal	...	×	× × ×	×	×		...	...	× × × × ×
Garganey	...	...	...	...	...		...		
Gadwall	×	× ×	× ×						
Pintail	×	× ×	× ×	×			...		× × ×
Wigeon	...	× ×	× ×	×			...		× × × ×
Pochard	×	× ×	×	...			...		× ×
Tufted Duck	...	× ×		...			...		× × ×
Scaup	×	×	× ×	×			...		× × ×
Long-tailed Duck ...	×	× ×	×	...	...		...		
Common Scoter ...		×	× ×	× ×	...		...		
Velvet Scoter		×	×	×	...		...		
Golden-eye		×	× ×	...			...		
Goosander		× ×	×	×			...		
Smew		×					...		
Ring Dove		× ×	× ×	×	...	...	...		×
Turtle Dove ...		...	...	...	...	× ×			
Quail	...	...	...	...	...	× × ×			
Great Bustard	×	×	×						
Little Bustard	×		×						
Land Rail	...			...	...				
Spotted Crake	...			...	...	× × ×			
Stone Curlew	...			...	...	× × ×			
Ringed Plover	...			...	...	... ...	...		×

S

Species	To N.; N.E.; E. — Departures. Feb.	March.	April.	May.	June.	From South to North. — Arrivals. Feb.	March.	April.	May.	June.	From South to North. — Coasting. Feb.	March.	April.	May.	June.
Kentish Plover	...	...	...	...	...	...		× ×	×						
Dotterel	...	...	...	...		...		× ×	×		...	...			×
Gray Plover	...	×	×	× ×		...		...	...	...	...	×	×		×
Golden Plover	×	× ×	× × ×	×	×	...		...	...	...	×	× ×	× × ×		×
Lapwing		× ×	× ×	×	×	...		...	...	...	×	× ×	× ×		×
Avocet		...	...	...	...	...		×	×						
Turnstone		×	×	× ×	×	...		...	...	...	...	×			
Gray Phalarope ...		...	...	...	...	...		...	...	...	×				
Red-necked Phalarope	...	...	...	...	...	...			× × ×						
Woodcock	×	×	× × ×	×		...		...	...	...		× ×	× ×		
Common Snipe ...	×	× × ×	×	×	...	...		...	...	...		×	×		
Jack Snipe	×	× × ×	× ×	×	...	...		...	...	...			×	×	
Ruff		...	...	...	...	...		×	× ×	×			×	×	×
Common Sandpiper		...	×	...				× ×	× × ×	...			× ×		
Wood Sandpiper ...		...	...	...				×	×						
Green Sandpiper ...		...	...	...				×	×						
Dusky Redshank ...		...	...	...			...	...	...					×	
Greenshank	...	...	...	...		...	×	×	× ×						
Common Redshank ...	×	× ×	× ×	×		...	...	...							
Curlew	× ×	× ×	× ×	× ×		...	...	...	...				×		
Whimbrel	...		...			...	...	× ×	× × ×	×			× ×		
Black-tailed Godwit			...			...		× ×	× × ×					×	
Bar-tailed Godwit			×										×	×	
Dunlin ...										...				× ×	

Little Stint	...		...	...	...	...		...	...	...	...		×	×	
Temminck's Stint ...	...		...	...	...	...		...	...	...	...		×		
Curlew Sandpiper ...	...		...	...	...	...		...	...	...	...		×	×	×
Purple Sandpiper ...	...		×	× ×	...	...		...	...	...	...		×	× ×	
Knot	...		×	× ×	×	...		...	...	...	...		×	× ×	×
Sanderling	...		× ×	× ×	×	...		...	...	...	...			× ×	×
Sandwich Tern ...	...		...	...	...	...		× ×	× × ×						
Roseate Tern	...		...	...	...	...		×	×						
Common Tern	...		...	...	...	...		×	× × ×						
Arctic Tern	...		...	...	...	...			× × ×				×	× ×	
Lesser Tern	...		...	...	...	...		×	× × ×						
Little Gull															
Glaucus Gull					...		...	...		...		×			
Buffon's Skua					...		...	...		...		×	×	×	
Pomatorhine Skua ...					...		...	...		...			×		
Richardson's Skua ...	...				...		×	×		...		×	×		
Puffin	...				...		×	×		...		×	×		
Little Auk	× ×														
Great Northern Diver	×		×	×	...					...					
Black-throated Diver	×	× ×	×	×											
Red-throated Diver ...	×	× ×	×	×	...					...	×	×		×	
Red-necked Grebe ...	×	×													
Sclavonian Grebe ...	×	×	...	...	...				...		×	×			

CHAPTER X.

THE AUTUMN ASPECTS OF MIGRATION IN THE BRITISH AREA.

Migration more Apparent in Autumn than in Spring—Difficulties of Observing the Phenomenon—Commencement of Autumn Migration in the British Islands—Arrival of Birds from the North and North-east—The Species that are the Earliest to Arrive—Growing Intensity of the Movement—The Hardier Species are Latest to Appear—Autumn Migration of Fieldfare and Redwing—The Earliest Departures from the British Area—Early Migrants Abnormal—The Growing Intensity of Southern Migration as Autumn advances—Duration of Migration Periods—The Migration into the British Area from the East—General Aspects of the Phenomenon—Abnormal Lines of Migration in Autumn—Cross Migration in Autumn—Reversal of Route by Migratory Birds—Erroneous Interpretation of the Facts—The True Explanation—Duration of Autumn Migration—Vertical Migration in Autumn—Order of Migration—Table indicating the Autumn Migration of Birds across the British Islands.

THE Spring Migration of birds across the British Islands has scarcely ceased, before signs of the autumn passage are visible. The Autumn Migration of birds from, to, and across the British Archipelago is even more interesting than the movement which is characteristic of the spring. Autumn migration is more palpable because the individuals engaged in it are so much more numerous, the old birds being accompanied by the young. This grand autumn movement, however, is most conspicuous amongst species that either winter in the British Islands or pass over them to more southerly latitudes. Among our common Summer Migrants, it is much more difficult to detect migration; they disappear one after the other from their accustomed haunts, and rarely can we give the exact moment of their departure. Owing also to the impossibility of distinguishing between individuals, we cannot readily mark the passage south of a portion of any species that only crosses our area on migration. Sometimes, however, as I have repeatedly observed, our regular breeding individuals will all disappear, and perhaps a week or so after the species will again frequent the district, this time represented by coasting migrants. These remarks apply most closely to inland migration; along the coast the autumn passage is much more distinct and noticeable. We remark precisely the same difficulties in spring.

The autumn migration of birds begins to be apparent in an average season in July. As we might also naturally expect, the first signs of the great Southern Flight are given by birds that breed in the highest latitudes, or colder north-eastern continental areas. These species or individuals experience the advent of winter much sooner than others inhabiting more southern and western areas. Autumn migration, therefore, begins not only from the north, but from the north-east. Among the first Passerine birds to make their appearance in our area in July are the Wheatear, the Swallow, the House Martin, the Pied Wagtail, the Song Thrush, the Robin, the Goldcrest, the Wren, and the Starling; among other birds the Cuckoo, the Corn Crake, and the Dotterel. All the birds that range far north or north-east, in fact, show signs of autumn migration in July. By the middle of the month flocks of northern waders begin to appear upon the coasts, mostly composed of young, but with a few old birds intermixed. As August comes on, the migration of all these northern and north-eastern species becomes decidedly stronger, reaching its climax perhaps during the latter half of that month and the first half of September; and, generally speaking, dying down again towards the end of the latter month. Among these migrants must be included many individuals of species that remain in our islands during the winter; many others, however, simply pass them as coasting migrants. The hardier species are later to arrive. Thus no migration into our area of the Brambling or the Fieldfare is apparent until September, and even then the passage is slight, not assuming large proportions until October, and even November. The Redwings' migration is apparent in August, stronger in September, and attains its maximum in October; but as we know this Thrush is more of an insect feeder, and its food is affected by frosts much sooner than that of the hardier berry- and seed-eating Fieldfare and Brambling. Coasting migration among the Charadriidæ, and the Terns and Skuas, is strongest during September and October; but amongst the Ducks and Geese it is at its maximum in October and November.

Although there is abundant evidence that migration into or across the British Islands begins in July, there is little or no evidence to suggest that a southern movement begins at all strongly during that month amongst individuals that breed in them. There is some evidence to suggest that a few individuals of such species as Spotted Flycatchers, Whitethroats, Willow Wrens, and

Turtle Doves are on the move during July. As none of these birds have extended their area north or north-east across Britain to continental areas or outlying islands, we can come to no other conclusion than that they are indigenous individuals moving south from our area. These very early migrants are probably birds whose broods have been destroyed or that have not been breeding at all, just as we know the earliest travellers from other areas are of a similar character. In August, however, the departures commence very generally, the migration assuming its greatest strength in September, and dying almost entirely away in October, but little being apparent in November. As the autumn commences we have abundant signs of the approaching departure of the British Summer Migrants. Nearly all are moulting, songs have ceased, social and gregarious tendencies are becoming more and more apparent. During August the Swifts and Cuckoos depart south, the migration continuing into September. Swallows and Martins, most of the young now strong upon the wing, significantly gather at well-recognized meeting-places preparatory to departure. Flocks of Terns, chiefly young, are moving south along the coast-lines. At first the autumn migration is remarkable for the preponderance of young; later the old birds are in the majority. Slowly as the mellow autumn days creep on, bird after bird disappears from the old familiar haunts; species after species takes its departure along the well-known routes to the south. Now and then a general rush of one or two particular species will be remarked. No particular hour seems chosen, the migration progresses day and night pretty evenly, so long as weather and wind are favourable. In the British

Islands the migration of those species that not only breed in them but pass them is of the longest duration; the migration of such species that are confined to our islands with no ancient lines of emigration across them is shorter. With the former species individuals continue to pass right through the autumn, the flight beginning and ending in a very gradual manner; with the latter species, once the movement south commences, it progresses steadily until the end, usually finishing as abruptly as it began. Thus the migration of the Wheatear, a species that has emigrated or extended its range across the British Islands not only to Scandinavia, but to Iceland and Greenland, extends over a period of five months, beginning in July, and actually not ceasing before November! On the other hand, the migration of the Reed Warbler, a species that never emigrated across the British Islands and whose range in them is limited, does not extend over more than a month, beginning say at the middle of August and closing by the middle of September or thereabouts. The migration of the Red-necked Phalarope is remarkably contracted both in spring and autumn; so also is that of the Garganey.

Up to the end of September, the general trend of migration across or to the British Islands is from the North or North-east. After that date a very perceptible change in the general trend takes place, and the predominating line of Flight falls nearly to due East. This is the first sign of that gradually approaching wave of migration from the east, from those continental areas which were colonized by birds whose emigrations may be said to have their base in the British Area. Until the middle of September this migration from eastern continental areas is more or less of a desultory character. During the latter half of that month it suddenly assumes a stronger aspect, culminating in a grand and mighty influx of birds, young predominating, lasting almost incessantly for perhaps a fortnight; then a lull occurs for a week or so; then another grand wave of not quite the same magnitude and duration, adults predominating, breaks upon our eastern sea-board, spreading perceptibly right across England to Ireland; after which the great flight is spent, resuming only in a fitful manner or entirely ceasing, as much of Eastern Europe and Western Asia become drained of most of their hardiest non-insectivorous birds. The birds that chiefly partake in this late migration from eastern areas are tabulated on a previous page (*conf.* p. 132). We have already dwelt at some length on the origin of this movement. A few remarks on its general character will now be all that is required. So far as numbers are concerned this eastern passage is the most important migration that breaks upon the British coasts in autumn. The number of species normally is not great. Night and day the steady inrush of migrants is constant and prodigious. For weeks birds may be remarked pouring into our islands by day and by night, or by day or night alone, the migration of each particular species varying considerably from year to year, sometimes being completed in a few weeks, sometimes continuing over as many months. Three of the most remarkable species performing this east to west migration in autumn are the Hooded Crow, the Goldcrest, and the Sky Lark; we might also add the Starling. For days and days together, sometimes, a nearly constant stream pours into the British Islands from the east. In 1882 it was computed that the migration of the Goldcrest into our area extended over a period of ninety-two days, commencing in August! It must, however, be remarked that the earlier arrivals were from the north-east, and that in computing the duration of the passage of this tiny species this fact is almost invariably overlooked. None the less remarkable are the autumn passages of Sky Larks and Starlings—vast waves of avian life that only spend themselves in the remote western areas of Ireland! I have already given many instances of this grand migratory movement in autumn in the *Migration of Birds* (pp. 255-258), to which volume I would refer the reader anxious for greater details.

It has been said that this East to West migration in autumn sometimes approaches us from points south of east. If such is actually the case the movement is entirely abnormal, as no bird whatever migrates in a northerly direction in autumn. Adverse winds or bad weather may have driven these migrants a little south of their normal course, but the reader may rest assured, if the passage is being undertaken with suitable wind and weather, the birds will never appear by any chance from points south of east, nor in a very marked manner from points north of east. The normal trend of this movement is east.

There is also just the same cross migration in progress in autumn as we have already found to be the case in spring. Birds coasting south across the British Islands pass almost at right angles the stream that is pouring in from the east across the North Sea. Every movement that is observed in spring is again repeated in autumn, only the directions are exactly reversed. While all this stream of migration from the east is in progress there is another movement going on amongst the same species to a great extent. These are the individuals that breed in our islands and pass to more southern areas to winter. The cause of this double movement has already been dwelt upon.

It has been asserted that some species reverse their routes almost entirely in autumn, and travel south by quite a different fly-line from that which they followed north in spring. Unfortunately, with too great a respect for the opinions of other naturalists, I myself have alluded to this movement in the *Migration of Birds* as though it were a fact. There cannot be the slightest doubt that this change of route is purely imaginary, for if we look closely into the facts they will be found to admit of a very different construction. When I wrote that volume I regret that the Law of Dispersal which I have attempted to explain and illustrate in the present work was then entirely unknown to me. I had accepted the general belief that a glacial epoch could cause southern emigration, and I was also labouring under the very general, if quite erroneous, idea that species were driven this way and that across the world without any governing impulse. Let us deal with the few instances known to me, and which I gave as examples of route reversal. The first species was the Nightingale. As a proof that this bird travels by a different route in autumn from that which it traverses in spring, it is shown that it passes Heligoland in April and May, but has never been caught there in autumn. Now Heligoland is situated at or near the very northern limits of this bird's distribution. There is nothing then very remarkable about a few birds overshooting the mark in spring and visiting the island, just as we know many southern birds visit our islands at that season under precisely similar circumstances. It would therefore be a most extraordinary thing indeed if this bird visited Heligoland in autumn, for to reach that island from its present normal limits of distribution it would absolutely have to take a *northern flight!* Again, the Dotterel is said not to visit Malta in spring but to pass regularly enough in autumn. Now the Dotterel migrates very rapidly in spring, its migration not lasting much more than a month; in autumn, however, as is customary with many species, it migrates more leisurely, the passage extending over four months, and in fact finds time to visit places it had to pass very quickly on its flight north. The Turtle Dove is commoner in spring at Heligoland than in autumn; the Whimbrel passes the British Islands in greater numbers presumably in spring than in autumn. But it has been remarked that the latter bird flies much higher in autumn, and consequently is disposed to alight less. The same remarks will probably apply to the Turtle Dove. In the latter cases, it will be remarked, a complete reversion of route has not been suggested. We might with equal propriety say that species of which individuals abnormally appear in our islands only in autumn or spring change their route according to season. The Little Bunting, the Rock Thrush, and the Yellow-browed Warbler are cases in point.

The migration of birds in autumn over, from, and to the British Islands extends over a period of quite five months. It begins in July and continues to November or even into the early part of December. Generally speaking, the tendency of migration in autumn is leisurely and more prolonged than in spring. The great bulk of autumn migration takes place in September and

October; the movement in every direction has a *southern* trend.

A short allusion to Vertical Migration in autumn will bring the present chapter to a close. The descent of mountain species from their upland haunts is just as characteristic of the autumn migration of birds as that migration which takes place in a latitudinal direction. The duration of the period extends over about three months, namely, from August to October, and is therefore much shorter in duration than the autumn migration of species elsewhere. This is probably due to the more rapid seasonal changes on mountains than on the lowlands, more of that sudden nature which marks the coming on of winter in the Arctic regions. Among the first birds to leave their mountain summer haunts are the Merlin, the Linnet, the Twite, the Lesser Redpole, the Pied Wagtail, the Golden Plover, the Lapwing, the Curlew, and the Dunlin. In all these species, however, the migration is only slight in August, and attains its greatest strength in September, a slight movement extending into October. The later birds to leave are the Stonechat, the Gray Wagtail, the Meadow Pipit, the Wood Lark, and the Sky Lark. With these species the migration is only slight during September, and reaches its maximum in October, one species, the Sky Lark, prolonging its flight into November. In our latitudes vertical migration is practically over by the end of October, but in more northern areas it has ceased for the year months earlier.

The following table will indicate the duration of the principal Migration across the British Archipelago in Autumn.

Species.	To S.; S.W.; W. — Arrivals. July.	August.	Sept.	Oct.	Nov.	From North to South. — Departures. July.	August.	Sept.	Oct.	Nov.	From North to South. — Coasting. July.	August.	Sept.	Oct.	Nov.
Missel Thrush			× ×	× × ×	× ×			× ×	× × ×	× ×					
Song Thrush			× ×	× × ×	×			× ×	× × ×	× ×					
Redwing			× ×	× × ×	×			...	...	...			×	×	×
Fieldfare				× × ×	× × ×			...	...	...				× ×	×
Blackbird			× ×	× × ×	× × ×										
Ring Ouzel			...	...	...			×	× ×				×	× ×	× ×
Black Redstart			× ×	× ×	×										
Common Redstart ...			...	...			× ×	× × ×							
Whinchat			...	...	...		× ×	× × ×	×	.			× × ×	×	
Stonechat			× ×	× ×											
Wheatear		...	...	...	...	×	× ×	× × ×	× ×	×			×	× ×	
Robin		× ×	× × ×	× × ×	×										
Nightingale		...	...	...	...	...	× ×	× ×							
Hedge Accentor ...		×	× ×	× × ×	×										
Whitethroat			...	...	...	×	× ×	× × ×	× ×						
Lesser Whitethroat ...		...	...	...	...		× ×	× × ×	×						
Blackcap		...	...	...	...	...	× ×	× × ×	×						
Garden Warbler ...		...	...	...	...	...	× × ×	× ×							
Chiffchaff		...	...	...	...	...	× ×	× × ×	× ×	×					
Willow Wren ...		...	...	...		×	× ×	× × ×	× ×	×					
Wood Wren ...		...	...	...	...	...	× ×	× × ×	×						
Reed Warbler ...		...	...	...	...	...	× × ×	×							
Marsh Warbler		...	...	...	...	...	× × ×	× ×							
Sedge Warbler ...		...	...	...	...	...	× ×	× × ×	×						

Grasshopper Warbler	...	...	...	...	...		× ×	× × ×							
Goldcrest	×	×	× ×	× × ×	× ×				...	...		× ×	× ×	× ×	×
Long-tailed Titmouse	×	×	×	× × ×											
Great Titmouse ...	×	×	× ×	× × ×	×										
Coal Titmouse	×	×	× ×	× ×	×										
Blue Titmouse	×	...	× ×	× × ×	×										
Wren	×	×	× × ×	× × ×	× ×										
Pied Wagtail	×	× ×	× × ×	× ×	×	...	× ×	× × ×	×						
White Wagtail	...	...	...	...	...	...	× ×	×	×	...				×	
Yellow Wagtail ...	...	...	...	...	...	...	× ×	× × ×	× ×						
Gray Wagtail	...	...	...	...	...	...	×	× × ×	× × ×						
Tree Pipit	...	...	...	...	...	...	× ×	× × ×	× ×	×					
Golden Oriole	...	...	...	...	...	...	...	...	×						
Great Gray Shrike ...	...	×	× ×	× ×	...	...	...	...	...	...		×	× ×	× ×	
Red-backed Shrike ...	...	...	...	...	...	...	× ×	× × ×	×	×					
Spotted Flycatcher ...	...	...	...	...	...	×	× ×	× × ×							
Pied Flycatcher ...	...	...	...	...	...		× ×	× × ×	×						
Swallow	...	...	...	...	...	×	×	× × × ×	× ×	×	×	× ×	× ×	× × ×	×
Martin	...	...	...	...	...	×	×	× × × ×	× ×	×	×	× ×	× ×	× × ×	×
Sand Martin	...	...	...	...	...	...	×	× × × ×	×	×					
Waxwing	...	...	...	×	× ×										
Goldfinch	...	...	...	× × ×	× ×										
Greenfinch	×	×	× ×	× × ×	× ×	...				...	×	×	× ×	× × ×	× ×
Linnet	×	×	× × ×	× × ×	× ×										
Chaffinch	...	×	× × ×	× × ×	× ×										
Brambling	...	...	×	× × ×	× ×	...				...			× ×	× × ×	×
Tree Sparrow	×	×	× ×	× × ×	× × ×	...				...				× ×	×
Yellow Bunting ...	...	×	×	× × ×	× ×		×	×	× × ×	× ×					
Corn Bunting	...	...	×	× × ×	× ×		...	×	× × ×	× ×					
Reed Bunting	...	...	× ×	× × ×	× ×		...	× ×	× × ×	× ×					
Snow Bunting	...	...	× ×	× × ×	× ×		...	...	...	...	...	...	×	× × ×	× × ×
Sky Lark	×	×	× ×	× × ×	× × ×		...	× ×	× × ×	× ×	×	×	× ×	× × ×	× × ×

SPECIES.	To S.; S.W.; W. ARRIVALS.					From North to South. DEPARTURES.					From North to South. COASTING.				
	JULY.	AUGUST.	SEPT.	OCT.	NOV.	JULY.	AUGUST.	SEPT.	OCT.	NOV.	JULY.	AUGUST.	SEPT.	OCT.	NOV.
Shore Lark	...	...	...	× ×	× ×		...	...	...	...	...	...	...	× ×	× ×
Starling	× ×	×	× ×	× × ×	× × ×		×	× ×		× ×	× ×	×	× ×	× × ×	× ×
Jackdaw		...	...	× × ×	× ×	...	...	...	...	...	...	...	...	× × ×	×
Carrion Crow		...	×	× × ×	× ×	...	...	×	× × ×	× ×					
Hooded Crow		...	×	× × ×	× ×	...	...	...	...	...			×	× × ×	× ×
Rook		×	×	× × ×	× ×	...	×	×	× × ×	× ×					
Jay			×	× × ×	× ×										
Swift			...	...	...	...	× × ×	× ×	×	×					
Nightjar ...			...	...		...	×	× × ×	×	×					
Wryneck ...			...	...	...	...	× ×	× × ×	×						
Hoopoe ...			...	...	...	...	× ×	× × ×	×						
Cuckoo ...			...	...	...	×	× ×	× × × ×	...	...	×	× ×	× × × ×		
Kingfisher			× ×	× ×											
Short-eared Owl ...			× ×	× × ×	×							×		× × ×	×
Snowy Owl		...	...	×	×										
Sparrow Hawk ...		×	× ×	× ×											
Rough-legged Buzzard			×	× ×	× ×				...	...	...		×	× ×	× ×
Hen Harrier			...	...					× ×	...	...		×	× ×	
Montagu's Harrier ...			...	...					× ×						
Hobby			...	...	...		...								
Merlin			...	...	...		×	× ×		...	...	×	× ×	×	
Honey Buzzard ...			...	...	...		...	× × ×							
Osprey			...	...	...		×	× ×	× ×	×					
Bittern			...	×											

White-fronted Goose	...	...	...	× ×	× ×	...	...	...		...				× ×	× ×
Bean Goose	...	...	...	× × ×	× ×										
Pink-footed Goose ...	...	...	×	× × ×	× ×										
Bernacle Goose ...	...	×	×	× × ×	× × ×										
Brent Goose	×	×	× ×	× × ×	× × ×	...		...		...	×	×	× ×	× × ×	× × ×
Hooper Swan	...	×	×	× ×	× ×	...		...		...	...	×	×	× ×	× ×
Bewick's Swan	...	..	...	×	×	...		...		...	...	...	...	×	×
Mute Swan	...	...	...	×	×										
Mallard	×	×	× ×	× × ×	× × ×			...	...	...	×	×	× ×	× × ×	× × ×
Teal	...	×	× ×	× × ×	× ×			...	...	...	...	×	× ×	× × ×	× ×
Garganey	...	...	...	...	...			×	×						
Gadwall	...	...	...	× ×	× ×										
Pintail	...	...	×	× ×	× ×				...	...	...	...	×	× ×	× ×
Wigeon	...	×	× ×	× × ×	× ×				...	...	...	×	× ×	× × ×	× ×
Pochard	...	...	×	× ×	×				...	...	...	...	×	× ×	×
Tufted Duck	...	...	×	× ×	×				...	...	...	...	×	× ×	×
Scaup	...	×	×	× ×	× ×				...	...	...	×	×	× ×	× ×
Long-tailed Duck ...	...	...	...	× × ×	× ×	...			...	...	...	...	...	× × ×	× ×
Common Scoter ...	×	×	× ×	× × ×	× ×	...			...	...		×	× ×	× × ×	× ×
Velvet Scoter	...	×	×	×	×	...			...	...		×	×	×	×
Golden-eye	...	...	×	× ×	× ×	...			...	...		...	×	× ×	× ×
Goosander	...	...	...	× ×	× ×	...			...	...		...	...	× ×	× ×
Smew	...	...	...	×	×										
Ring Dove	...	...	×	× × ×	× × ×			...	...	...	...		×	× × ×	× × ×
Turtle Dove	...	...	...	...	...			× × ×	×						
Quail	...	...		...	...			× × ×	× ×						
Great Bustard	...	...		×	×										
Little Bustard	...	...		×	×										
Land Rail	...	...		...	..		×	× × ×	× ×	×					
Spotted Crake	...	...			...			×	×						
Stone Curlew	...	...							× × ×						
Ringed Plover	...														

SPECIES.	To S.; S.W.; W. ARRIVALS.					From North to South. DEPARTURES.					From North to South. COASTING.				
	JULY.	AUGUST.	SEPT.	OCT.	NOV.	JULY.	AUGUST.	SEPT.	OCT.	NOV.	JULY.	AUGUST.	SEPT.	OCT.	NOV.
Kentish Plover ...	...	...	...	...	...	...	× ×	× ×	×	×					
Dotterel	...	...	...	...	...	×	×	× × ×	×		×	×	× × ×	×	
Gray Plover	×	× ×	× × ×	× ×	× × ×	...	...	...	...	...	×	× ×	× × ×	× ×	× × ×
Golden Plover ...		× ×	× × ×	× ×	× × ×	...	...	...	...	...	×	× ×	× × ×	× ×	× × ×
Lapwing		× ×	× × ×	× ×	× × ×	...	...	...	...	...	×	× ×	× × ×	× ×	× × ×
Avocet		...	...	...	...	...	×	×							
Turnstone		×		×	×	...	...	...	...	...	×	× ×	× ×	×	× ×
Gray Phalarope ...		×	×	× ×	×	...	...	...	...	...	...	×	× ×	× ×	×
Red-necked Phalarope		...	...	...	...	...	...	× ×	× ×						
Woodcock		×	×	× ×	× × ×	...	...	...	...		...	×	×	× × ×	× × ×
Common Snipe ...			× ×	× × ×	× × ×	...	...	...	...	...	...	×	× ×	× × ×	× × ×
Jack Snipe		...	×	× × ×	× × ×	...	...	...	...	...	...	...	×	× × ×	× × ×
Ruff		×	× × ×	×	×	...	...	...	...	...	...	×	× × ×	×	×
Common Sandpiper ...		...	...	...	...	×	× ×	× ×	× × ×	...	×	× ×	× ×	× × ×	
Wood Sandpiper ...		...	...	...	...	...	×	×							
Green Sandpiper		...	...	...	...	...	×	×							
Dusky Redshank ...	...	...	...	...	...	...	...	...	...	...		×	× ×		
Greenshank	...	...	...	...	...	×	×	× × ×	× × ×						
Common Redshank ...	×	× ×	× ×	× ×	× ×	...	...		...		×	× ×	× ×	× ×	× ×
Curlew	×	× ×	× × ×	× ×	× ×	...	...	...	...	...	×	× ×	× × ×	× ×	× ×
Whimbrel		..	...	...		×	× ×	× ×	×	...	×	× ×	× ×	×	
Black-tailed Godwit ..	...	...	...	...	...	×		× × ×	× ×	...	×	×	× × ×	× ×	
Bar-tailed Godwit ...	×	×	× × ×	× × ×	×	...		...	...	...	×	×	× × ×	× × ×	×
Dunlin	...	× ×	× × ×	× ×	...	...		...	...	...	...	× ×	× × ×	× ×	

Little Stint	...	...	...	...	...	...	...		...	...	
Temminck's Stint ...	...	...	...	...	...	...	...	...	...	...	
Curlew Sandpiper ...	...	...	...	...	...	...	...	...	...	...	
Purple Sandpiper ...	...	...	× ×	× × ×	× ×	...	...	...	...	...	...
Knot	×	× ×	× × ×	× × ×	× ×	...	...	...	...	...	×
Sanderling		× ×	× × ×	× × ×	×	...	...	...	...	...	×
Sandwich Tern ...		...	...	...	...	...	× ×	× × ×	×		
Roseate Tern		...	...	...	...	...	×	×	×		
Common Tern		...	...	...	...	...	×	× × ×	× ×		
Arctic Tern	...	...	...	...	...	...	×	× × ×	× ×	...	
Lesser Tern	...	...	...	...	...	...	× ×	× ×	×		
Little Gull	×	×	× ×	× ×	×	...			...	...	×
Glaucous Gull	...	...	×	×	×	...			...	...	
Buffon's Skua	...	...		...	...	...				...	
Pomatorhine Skua ...	...	...		...	...	...		...		...	
Richardson's Skua ...	...	...		...		...		× ×		...	
Puffin	...	...		...	...	...		×	×	...	
Little Auk	...	...		×	×						
Great Northern Diver	...	...		× ×	× ×						
Black-throated Diver	...	...		...	×						
Red-throated Diver ...	...	...			×						
Red-necked Grebe ...				×	×						
Sclavonian Grebe ...	...			×	×						

CHAPTER XI.

INTERNAL MIGRATIONS AND LOCAL MOVEMENTS IN THE BRITISH ARCHIPELAGO.

Meagreness of Data bearing on this Question—Local Movement in Spring and Summer—Internal Migration always takes place within Normal Areas of Dispersal—Birds do not Wander from their Areas—Comparative Movements of the Snow Bunting, the Northern Bullfinch, and the Crested Titmouse—Emigration only Undertaken during the Season of Reproduction—Local Movement amenable to Law—Effects of Severe Winters on Birds—Results of such Movements ineffectual in extending Area—Redwings and Severe Weather—The Want of carefully-kept Records—Local Movements at Lighthouses—Irruptic Movements—Their Futility as Colonizing Agents.

WE cannot well dismiss the subject of Migration in the British Islands without some brief allusion to all those local movements of birds that take place within that area. The subject is a much more difficult one than might be suspected. Unfortunately at present the data are too meagre to allow of much accurate generalization or deduction. A vast amount of observation has been made, but the haphazard way in which the facts have been collected is a very serious detraction from the value of such observation. Notwithstanding the apparent fortuitous character of much of this internal migration and local movement, I am compelled to believe that it is governed by law.

In the British Islands these local migrations apparently are only undertaken in winter. I say apparently, because it is by no means proved that there is not a considerable amount of local movement taking place at other seasons and entirely overlooked. In winter birds are more gregarious, more easily observed, and the initiating causes of local migration, such as rises and falls in temperature and storms, are readily and easily defined. In spring and summer much of this local movement might go on, as indeed I strongly suspect it does, without being noticed. Birds both in winter and summer follow their food, the general area of their distribution is that in which they can find sustenance, and as this food-supply varies a good deal according to season, both in the matter of description and locality, it is only natural that birds should undertake some local journeys in quest of it. So far as I can ascertain, *all this internal migration and local movement takes place within the normal area of dispersal of every species.* There is no evidence whatever to show that a species will wander from its normal area in quest of food, and the laws which govern its dispersal inexorably confine it to that area, and the species will perish therein if the conditions change and render existence impossible. As I have previously shown, species never extend the area of their dispersal in winter, emigration or range expansion can only take place in summer or just prior to the season of reproduction. It must also be remarked, that not a single species taking part in these local movements in winter in our area is non-indigenous to that area. No matter how severe the winter may be on the Continent, the Pine Grosbeak, the Nutcracker, the Crested Lark, or the Central Russian Blue Tit (*Parus pleskii*) will never normally cross the North Sea in quest of food in company with the vast flights of Starlings and Sky Larks, and other well-known indigenous species that we know even in midwinter pass to and fro in response to falling temperature or storms. The Siberian Jay, an inhabitant of the pine forests of North Russia, never wanders to our area in quest of food, although the Hooded Crow which frequents the same forests pours into England to winter, and is known to cross over to us as late as December! The Snow Bunting is more or less on passage to and from our islands and the Continent all the winter through, but the Shore Lark is only an abnormal visitor on spring and autumn passage. The Northern Bullfinch (*Pyrrhula major*) never visits our area in winter in company with the hosts of Finches that cross the North Sea prompted by severe weather. The Crested Tit never crosses to us from the pine and oak forests of Germany and Holland, yet the Blue Tit does so in considerable numbers; the emigration of the former species was entirely continental, that of the latter partly continental and partly from a British base across the North Sea plains. What I want therefore to impress upon the reader is that no matter how extensive this winter movement may be, no matter how birds may wander to and fro in response to variations of weather, such journeys all take place within the normal area of distribution of such species, and within the limits of their usual migrations. These movements are not in any sense analogous to what is presumed to have taken place in a Northern Glacial Epoch—that utterly fictitious southern emigration of life—they are purely local, and can never extend beyond the geographical limits of the species partaking in them. *The Law of their dispersal, which forbids southern extension of range either in summer or winter, and only permits northern extension or emigration at all during or just prior to the season of reproduction, is inexorable and immutable.*

Having thus placed the subject on what I believe to be a thoroughly sound and satisfactory basis, we will pass to a more detailed study of the phenomenon. If we admit that the whole phenomenon of Local Movement or Internal Migration is confined within certain limits and controlled by law to those limits, I think we must also admit that the individuals partaking in it conform to certain governing impulses. We have seen how closely species and individuals and their descendants are confined to certain routes, to certain limits, from which, broadly speaking, they seldom diverge. It is therefore difficult to believe that all this winter migration is entirely fortuitous. Birds, individuals, have certain routes, inhabit certain areas or districts to which they are attached. It seems probable, then, that the individuals partaking in these various winter migrations are moving to and fro, according to changes of weather in the area they occupy, along certain routes, which routes indicate the lines of past emigration or range expansion. Birds move south or north along a coast, or east and west across a sea, but there is a certain amount of method in the flight. Now most of this winter migration, across the North Sea for instance, takes place within a comparatively narrow area. No matter how vast may be the winter flights of the Snow Bunting, for instance, we never observe any of their effects in the extreme west of England. Birds of various species may literally swarm along our eastern coasts; all winter migrants from the Continent, yet the more inland districts are not affected. No matter how abundantly the Hooded Crow may pour into the districts of the Wash, we never see a corresponding increase of Hooded Crows say in the neighbourhood of Sheffield, not 80 miles away, and where the species is very rare. Wading birds, Ducks and Geese, may oscillate between the Continent and our eastern sea-board, but the movement is purely local, confined to the feeding grounds of those individuals of the species that frequent the districts affected. To all these local migrants the route must be familiar; but there is not the slightest trace of any attempt to increase or prolong that route into other areas. Birds may seem to be flying this way and that, entirely at the mercy of the elements, but there can be no doubt whatever they know perfectly well where they are going—they are following familiar routes to other and more open haunts, anticipating a storm perhaps by hours, or retreating from a frost that has suddenly sealed their feeding places. Spasmodic much of this winter migration may be, fitful as the meteorological changes that initiate it, but unquestionably in conformity with order and law.

Every observer of birds must have often remarked the effects of an unusually severe winter upon them. Species will then come near to houses or visit localities where they are never seen under ordinary circumstances. I have known Red Grouse, when the moors have long been covered with snow, resort to the farmyards, and even to villages and towns. Scores of similar instances might be given; and in some continental districts, where the weather has been far more severe than with us, still more extraordinary cases have occurred—of wild birds visiting civilized places to seek for food. Now in the first place let it be remarked that however unusual the locality may be in which such species may appear under these exceptional circumstances, it is always within the normal area occupied by that species. A Nutcracker will never come to an English cottage door for food, no more than a Robin will ever appear at the threshold of a Canadian settler. In the second place, the straying of a species from its accustomed haunts is purely abnormal —a struggle for life in fact of an individual—and such an action in the majority of cases would not save the species from extermination if it succeeded in saving that individual. The conditions for successful reproduction, found only in the normal haunts of the species, would be wanting, and the inevitable result would be a more or less rapid extinction throughout the area affected.

No matter how an area may abound with food in winter, it will not be visited normally by any species whose area of distribution is beyond it. Vast numbers of birds die during a severe winter in the British Islands alone, if food fails in the local area of distribution. I have known our flocks of Redwings, which used to come to certain localities and remain in them the winter through, be almost exterminated during an exceptionally severe season, and not to regain their usual numbers for years afterwards. And this, mind, in a locality where a flight of a very few miles would have averted the disaster.

But the birds and their descendants that wintered in my neighbourhood had done so for time out of mind; they came there by certain routes; they knew of no more southerly areas, and each time a severe season overtook them they were more or less decimated. A study of the various local movements and internal migrations in our area alone promises some very interesting results. We are often told that British ornithology is pretty well played out; here then is a new field for observation. We want a carefully-kept record not only of the general movements of the birds in a district, but of those that pass through it, combined with keen incessant observation as to the causes of such movements, their direction and duration, and the season of the year in which they are undertaken. Even a record of the summer and autumn wanderings of such a common species as the Sparrow will be of service. Of course these observations are entirely separate from the usual migrations (if any) of a species to and from any district.

Some very interesting instances of local movement have been observed at our lighthouses, especially along the eastern coast-lines. Others occur across the Irish Sea. Equally interesting movements have been remarked along the coasts. It is usually remarked that these influxes correlate with periods of severe weather on the Continent, or in the northern British areas. At Heligoland a great deal of local winter movement takes place. It is impossible in the present state of our knowledge to enter into greater details. We do not possess sufficient data to generalize or deduct very extensively. At present all we can say with certainty is, that this local movement is a fact; it has been observed over a great number of years, but the details of the migration still remain to be worked out. From what we have already learned of the emigration or dispersal of birds, we cannot class Internal Migration and Local Movement as fortuitous.

A few words on Irruptic Movements will bring the present chapter to a close. These movements are entirely abnormal, and are rarely if ever attended by success. One of the most remarkable instances of irruptic movement is that furnished by Pallas's Sand Grouse. Details of the several irruptions of this species into Western Europe (in every important invasion, however, be it remarked, to increase *breeding*, not *winter* area) need not be given here; they are doubtless fresh in the minds of most readers. The last great spasmodic invasion which may be said to have spread almost entirely over Europe took place in 1888. The chief point of interest to us, so far as the present subject is concerned, is the absolute failure of these birds to establish themselves in any portion of the area they invaded. True, many of the birds made more or less successful attempts to breed, and some individuals may have lingered on in their new home for several years; but I much doubt if a single Sand Grouse out of the thousands that invaded Europe in 1888 now survives. Similar irruptions of Jays and Rose-coloured Pastors have been remarked, but in no case has any permanent success followed the movement. These movements very forcibly demonstrate how futile irruptic colonization is, and how rarely such individuals enter new areas where conditions of life are favourable to them. They are utterly abnormal means of dispersal, and in the great majority of cases must inevitably end in failure. Some ornithologists are disposed to explain difficult problems of geographical dispersal by similar irruptic movements, but in the face of such damning proof to the contrary such an explanation should never be invoked, unless supported by absolute demonstration. Once more let me assert most emphatically that the dispersal of birds, nay of all organisms, is governed by Law, not by chance, that it is not fortuitous but the result of design.

Once more I repeat, Birds do not increase their range in winter. *It is this all-important fact that keeps species to their normal areas of dispersal.* If winter conditions led to extension of range or emigration, then species would wander fortuitously far and wide, and such a thing as geographical limits would be almost unknown —especially in temperate and boreal latitudes, where the food supply is ever a fluctuating one. The more I study the question, the more I am convinced that Dispersal or Range Expansion is solely the result of increase, and that the spasmodic, totally abnormal, wanderings of species in quest of sustenance during winter or the non-breeding season, can never lead to emigration or permanent extension of area.

CHAPTER XII.

SUMMARY AND CONCLUSION.

The Present Volume illustrates the Development and Application of a New Law of Dispersal—Past Geographical and Climatic Changes—The Glacial Epoch and its Bearing on the New Law of Dispersal—Effects of the Glacial Epoch on Species—Range Bases—Application of the Law of Dispersal to the Range Contraction and Emigration of British Birds—The Migration of British Birds—Routes of Migration—Conditions of Flight—The Spring and Autumn Aspects of Migration in the British Archipelago—Internal Migrations and Local Movements in the British Area—Irruptic Movements—The New Law of Dispersal—Its Bearing on the Arctic Element in South Temperate Floras—Inter-polar Floras—Impossibility of Emigration of Plants from North to South—Presence of Southern Genera in Europe—The Andes as a Route for the Southern Migration of Plants—The Floras of Mountains in the Torrid Zone during Pre-Glacial Ages—Arctic Floras could never have been Developed in the Polar Regions—Dispersal of Plants North and South from Equatorial Range Bases—Effects of Glacial Epoch on Northern Floras—Absence of many Species from Equatorial Range Bases—The "Retreat" of Plants a Myth—Conditions of Successful Dispersal—The Flora of the Mountains of Asia—Inter-hemisphere Species—Species in Polar and Temperate Zones—The Distribution of Plants in Africa—The Temperate Flora of South Africa—Northern Emigration from Antarctic Centres obviously Erroneous—The Bearing of this New Law of Dispersal on the Absence of Southern Types from the Northern Hemisphere—The Dominant Southern Flora—Its Dispersal from Range Bases South of the Equator—The Problem of Migration and

Geographical Dispersal hitherto attacked at the Wrong End —Exterminating Influence of Glacial Epochs—Powers of Organisms to Extend their Areas of Dispersal—This Dispersal not Fortuitous but governed by Law.

THE subject of the present volume, the Migration and Dispersal of British Birds, has been selected to illustrate the development and application of what I believe to be an entirely new Law governing the Distribution, Emigration, or Dispersal of Species. In elucidating this subject a very wide and varied series of phenomena have had to be dealt with, yet no more than were absolutely necessary to furnish a satisfactory and tolerably complete explanation of the facts. In order to render the whole subject as clear as possible, it may be advisable not only to recapitulate the most salient features in a concluding chapter, but to deal with a few facts bearing on this new Law of Dispersal that could not well have been introduced into the general subject matter of the volume.

In dealing with the Migration and Dispersal of British Birds, we found it impossible to make any progress until we had traced out the past geographical and climatic changes, not only in the British Area itself, but in more or less adjacent areas, during late Pliocene and throughout Pleistocene time. We had first to ascertain as correctly as the state of present knowledge admits, the condition and the physical aspects, not only of Europe, but of a great part of Africa, during those periods. Taking the present distribution of species as a guide, we ascertained that a great many of those changes of climate and of geographical conditions upon which astronomers, physicists, and geologists insist, were in absolute harmony with such dispersal. We have shown that the gradual severance of the British Islands from continental land is also quite in accord with the present distribution of birds over their area; whilst the necessity of a former much greater land extension between Greenland and Europe is shown to be imperative.

We next pass to a consideration of the Glacial Epoch, and its results upon the fauna and flora of the regions affected. By the aid of a new Law of Dispersal, I have endeavoured to show that the effects of this Glacial Epoch must have been very different from those which biologists have universally accepted and described. I have shown that the conditions of the Ice Age, instead of being grand incentives to Southern Emigration, exerted a vast exterminating influence, and that they must have caused the utter extinction of every species whose breeding range was entirely confined to the areas glaciated, or sufficiently within the influence of glaciation to render existence impossible. The effects of the Glacial Epoch on the dominant Euro-Asian fauna are shown to be exterminating rather than incentive to Southern Emigration. The only species that survived were those that occupied a southern and continuous range base during Pre-Glacial time. These southern Range Bases (or what are perhaps better described as Refuge Areas), so far as British birds are concerned, are then defined. I then proceed to show that the breeding range of all surviving species must have extended at least as far south as these limits during Pre-Glacial time, and that the Glacial Epoch exterminated the northern portions of such species, or contracted the range of such as were migratory, the habit of migration not being acquired during the Glacial Epoch, but a result of range extension from more southerly areas during favourable intervals of climate in either hemisphere. Such species were, and probably always had been, Inter-hemisphere or Inter-polar, with a more or less ancient Equatorial range base. An Inter-hemisphere species, for instance, like the Swallow, or an Inter-polar species like the Hudsonian Godwit, could never be exterminated by a Glacial Epoch. As the summers became colder, the breeding range would gradually become lower—the species probably suffering considerably meantime, owing to the increasing adverse conditions under which the young of the most northerly breeding birds would be reared, in some seasons perhaps none at all surviving—and sink back upon itself, until it was driven as far north or south as the winter quarters commenced, which represent the centre of dispersal or range base of the species. If the winter range or range base of a species did not extend south of the Equator the species would never emigrate south, but continue to occupy that area until a return of more favourable conditions for northern extension or range expansion again took place. If the winter range or range base of a species did not extend north of the Equator, the species would never emigrate north, but remain stationary until favourable conditions returned, and permitted a southern extension or range expansion. If the range base extended both north and south of the Equator, that Polar or Temperate area would be occupied by those Inter-hemisphere and Inter-polar species which presented the most favourable conditions for reproduction. At the present time these conditions are almost exclusively in the Northern Hemisphere, but during the last

Glacial Epoch in that hemisphere the Southern Hemisphere would most likely present the most favourable conditions for successful reproduction, if land extended southwards. Thus from a common equatorial range base during the course of ages, would the breeding area of these Inter-hemisphere and Inter-polar species be to a varying extent reversed in the former group, and completely reversed in the latter, the breeding grounds in one hemisphere becoming the winter quarters in the other (conf. *Migration of Birds*, pp. 149-152). On the other hand, Northern Hemisphere species living permanently in the northern areas, with no southern breeding range base beyond the limits or fatal influence of glaciation, would gradually be exterminated; or Southern Hemisphere species living permanently in the southern areas with no northern range base beyond the limits or fatal influence of glaciation would also be as surely exterminated, because the Law governing their dispersal forbids retreat. The significance of these facts, and their bearing on the Arctic element in South Temperate floras, I hope presently to show.

We then proceeded to apply this Law of Dispersal to the glacial range contraction and Post-Glacial emigration of British birds, and endeavoured to show how existing species were preserved during the Glacial Epoch; how such surviving forms emigrated or expanded their range north with the return of more favourable climatic conditions. Passing, then, from a study of universal emigration during remote ages to more local range expansion actually in progress at the present time, we proceeded to show that the movement is governed by precisely the same law that controlled it in the past. A chapter on Island Avifaunas, their origin, the conditions under which they are maintained, and their relation to glacial conditions, bring the first part of our subject to a close.

Having satisfactorily accounted for the presence of the British avifauna, we next proceed to a study of its Migration or Season Flight. Routes of Passage engage our attention first. We learn how gradually a Route of Migration has been formed—how slowly a change of climate might curtail it, or a return to more favourable conditions assist in its expansion—how species never extend their winter area, such expansion invariably being the result of an increase of breeding population—birds Emigrating and Migrating solely to breed! Then the various routes of migration to the British Islands are divided into classes, each being dealt with and described in turn. First we trace the paths of the Summer Migrants to that area, and show how the most important routes are in the closest vicinity to the continental land masses, the weakest migration taking place in the most westerly areas. The significant bearing of these facts upon the distribution of species within the British Archipelago is next discussed, and many anomalies of dispersal are satisfactorily explained by law. Passing on to the migration across the North and Irish Seas, we endeavour to trace the correlation of routes with submergence in those areas. We next deal with the routes of species that winter in our islands or pass over them as coasting migrants to other lands. A brief account of the inland continuation of these routes, showing how such were probably formed, brings that portion of the subject to a close.

Having shown the origin of these Routes of Migration, we next proceed to discuss the Conditions of Flight. Dealing first with the Instinctive impulse of Migration, we then pass to the question of how birds are able to traverse these routes with such apparent precision. Experience rather than Inherited Impulse is shown to be the guiding influence. The altitude of migration flight, and the order of migration are next discussed. The daily time of migration, and the gregariousness or otherwise of birds on passage are then described; whilst finally the perils of migration are briefly treated.

Our next two chapters are devoted to a general description of the Spring and Autumn Aspects of Migration in the British Area. We have traced the phenomenon in spring from its earliest beginning in the departure of those migrants to the east across the North Sea that breed in continental areas, and the almost simultaneous arrival of other migrants, many of the same species from more southern areas, that breed in Britain. Later we describe the departure north or north-east of our winter visitors; together with the vast amount of coasting migration that passes over the British Area, composed of birds that winter to the south of us and breed to the north. Coming then to the arrival of spring migrants to the British Islands, we trace the movement from its beginning onwards through the months that it continues, until it finally dies away for the season. A few remarks on the vertical migration of birds in spring, species that ascend to various elevations for the purpose of breeding, together with a table indicating the duration of Flight, bring the spring aspects of the phenomenon to a close. Entering then upon the Autumn Migration of birds, we trace the first signs of the movement by migrants entering our area from the highest and coldest latitudes as early as July. As the autumn advances the migration gains in strength, and our own summer birds begin to take their departure, that event being more or less directly preceded by the annual moult, and in many cases by the suggestive gathering together of individuals. The vast amount of coasting migration is described, as is also the order of passage and the duration of flight. In the late autumn a very noticeable change in the direction of migration is apparent, and we then proceed to deal with the vast east to west movement across the North Sea. Various apparent anomalies of Flight are then discussed, as are also the presumed change of route and the autumnal descent of migrants from their mountain breeding places. The subject will be too fresh in the reader's mind to require recapitulation in further detail.

A short chapter dealing with the various Internal Migrations and Local Movements of birds in the British Islands introduces us to a very important branch of the subject, which unfortunately cannot be treated in a very detailed manner owing to the utter lack of necessary information. We have shown, however, that all this Internal Migration is purely of a local character, absolutely confined to the areas occupied normally by species undertaking it, and therefore controlled by that Law of Dispersal which forbids extension of range during the season of non-reproduction. We next proceed to discuss the effects of abnormally severe winters on birds, and the impotency of such effects to increase or to transpose geographical area; bringing the subject to a close with a brief allusion to Irruptic Emigrations, showing their abnormal character and their failure in the majority of instances to extend distribution.

Among the more important points dealt with in the present volume may be mentioned the following:—

I. A new Law of Dispersal.

II. Polar Dispersal—from either Pole—a myth.

III. Glacial Epochs exterminated Life; did not cause Emigration or "retreat" from adverse climatic conditions.

IV. The sources or Range Bases whence the British Post-Glacial Avifauna has been derived.

V. Past geographical Mutations have been shown to be in harmony with the present geographical Distribution of Species.

VI. Endemic island species are never produced on routes of migration of closely-allied parental species.

VII. Emigration or Range-extension is rarely made across wide water areas; but Migration, once established across such, when the land was continuous or nearly so, is only arrested by extermination.

VIII. The origin of the West to East Migration across the North Sea.

IX. Additional light has also, I believe, been thrown on the Migration Routes of Birds.

X. Internal Migrations, Local Movement, Irruptic Emigration, and Recent Emigration have all been treated from what I believe to be new points of view.

XI. A short comprehensive account of the Spring and Autumn Aspects of Migration across the British Archipelago.

XII. The important bearing of the new Law of Dispersal on the Distribution of Floras.

XIII. The development of "Arctic" species.

XIV. Migration a result of Normal Increase, and not initiated by winter conditions, or retreat from adverse conditions.

The object of the present work has been to a great extent to demonstrate a hitherto undiscovered Law governing the dispersal of species. That Law forbids the southern emigration of Arctic and North Temperate forms to Antarctic and South Temperate latitudes. So far as I can ascertain, every biologist of note insists upon this Southern Emigration. The "arctic element in South Temperate floras," to quote Dr. Wallace, has been repeatedly brought forward in support of this view; and so generally has this interpretation of the facts been accepted by naturalists, that to question its truth seems little less than a rank biological heresy. If these facts have been correctly interpreted, then our Law of Dispersal cannot apply to floras; animals may bow submissive to its edicts, but plants may set it at defiance and demonstrate its impotency. I hope, however, presently to show that not only does the phenomenon of "Arctic" types in the Southern Hemisphere conform to this Law of Dispersal, but that it actually illustrates it in no uncertain way.

Sir Joseph Hooker graphically describes a "continuous current of vegetation" extending from Scandinavia to Tasmania; a second very similar current occurs along the mountain systems of North and South America; whilst a third stretches across the highlands of the African continent. This flora, which has been described as the "Scandinavian," is universally admitted by botanists to possess astonishing colonizing power, due perhaps to the exceptional means of dispersal with which the plants that compose it are endowed. Of the various means by which this dominant flora has emigrated from certain centres it is quite unnecessary here to speak; that does not concern the point at issue in the slightest degree. This flora, which is designated by the excessively inappropriate epithet of "Scandinavian," is in reality an Inter-polar Flora, and occupies a precisely analogous position to that of the Inter-polar avifauna, of which various species have from time to time been mentioned during the course of our investigations. It is an Inter-polar Flora with a well-established equatorial range base on the mountains and highlands of the torrid zone—a flora, therefore, which no glacial epoch at either pole could completely exterminate—a dominant flora that, notwithstanding the complete extermination which might overtake all those species or portions of species within the sphere of glacial influence, would still be preserved on its southern and equatorial bases, and emigrate north or south again from those bases towards whichever pole where glacial conditions were passing away. As we found to be the case with Inter-polar species of birds, so we also find with this Inter-polar flora that the Polar region best suited to the requirements of such avifauna or flora is that where it predominates. At the present time the Arctic regions are best adapted to the requirements of this Inter-polar flora, and consequently there it thrives best, is most abundant, and most widely dispersed. In the Antarctic region conditions are unfavourable, therefore this flora is neither dominant nor abundant, having been exterminated during the last glacial epoch at the Southern Pole, and only lingering on a few northern bases in the immediate vicinity of that area it once must have occupied so widely and in such abundance. To describe this flora as either "northern," "Arctic," or "Scandinavian," is therefore a most erroneous definition. We might, with equal propriety, speak of the "Arctic" element amongst birds, penetrating even to Patagonia, South Africa, Australia, and New Zealand, and even more remote latitudes; whereas, as we have already seen, such birds are Inter-polar and belong as much to the Antarctic as to the Arctic region.

There can, therefore, have been no emigration of plants from north to south. The range of the Polar flora has been contracted by extermination (how many times we may probably never know) as far as glacial influences have radiated from either pole; it has been expanded from such lower range bases as conditions favourable to Polar emigration have returned. We have precisely the same class of phenomena among plants as we have among birds—in many cases identical species in both hemispheres of the more Polar ranging types; distinct species generically identical, and southern representative forms, in the case of the more Temperate ranging species—both indicating equatorial centres of dispersal, with little or no isolation or discontinuous range area in the Inter-polar species, but with varying degrees of isolation and discontinuity in those we have already classed as Inter-hemisphere species. It has been said that some botanical genera now characteristic of the Southern Hemisphere appear to have been originally derived from Europe! Now in the first place some of our most eminent botanists have utterly discountenanced such statements; and in the second place, even admitting the identifications to be correct, there is no evidence whatever to disprove that these genera have not spread north and south respectively from an equatorial base—suffered extermination in Europe, but still survive in Australia. I might here take the opportunity of remarking, that in a great many cases these equatorial range bases must for obvious reasons have been entirely obliterated, as species have moved north and south from them, and become resident types in higher latitudes.

The Andes and the Rocky Mountains, stretching in one almost continuous line from the Arctic to the Antarctic regions, are stated by Dr. Wallace "to have formed the most effective agent in aiding the southward migration of the Arctic and North Temperate vegetation." Now, for the sake of argument, we must presume that the equatorial or torrid portions of this continuous mountain chain (and also by analogy of *all* other mountains and highlands within the torrid zone) held a flora of some kind, suited to the requirements of a very elevated cool climate, before any glacial epoch came on to drive these imaginary Arctic plants southwards. There is perfectly uncontrovertible evidence to prove that before glaciation the lands in North Polar latitudes contained a luxuriant flora, even of a semi-tropical character, and that therefore no "Arctic" or "Scandinavian" flora could have existed dominantly in those latitudes, if at all. If they did so exist, "*in no one case has a single example of such a fauna or flora been discovered of a date anterior to the last Glacial Epoch.*" (The italics are mine.) As the climate slowly changed and cold conditions came on, from where or whence did the "Arctic" flora arrive? True, many of the species—such as the dwarf birch for instance—are but modified forms of more Temperate types, and are the result of Post-Glacial invasion from southern bases; but, paradoxical as it may appear, the strictly Arctic or Inter-polar flora could never have been developed within the Polar regions of either hemisphere. We are forced to the conclusion that it must have had an origin in a region where cold conditions have existed unchanged for a vast and indefinite period of time. No part of the world fulfils these conditions except the mountains and highlands within the tropics. Here we can conceive how vegetation slowly crept up these heights from the equatorial plains, becoming modified as it reached those zones where cool climates prevailed —how it spread in strict accordance with the Law of Dispersal north or south along the mountain chains towards the poles, entering the Arctic or Antarctic regions, and becoming dominant as they offered favourable conditions for its increase, the altitudinal range becoming lower as the latitudinal range became higher. We can also understand how when a change to a warm climate occurred (a mild inter-glacial period) this flora dwindled away and perished in these Polar latitudes, and was only preserved on the higher and more southern range bases; or in like manner how when the Glacial Epoch assumed an intense phase it became exterminated and buried under the ice-sheets and snow-fields, but maintained its existence through the southern range bases which preserved those portions of the species that dwelt in areas beyond glacial influence, supplying fresh colonists to emigrate towards the glaciated regions, as soon as conditions became favourable. It is said by Dr. Wallace (referring to the Andes) that there are "between sixty and seventy northern genera in Fuegia and Southern Chile, while about forty of the species are absolutely identical with those of Europe and the Arctic regions"; and further, "as only a few of these species are now found along the line of migration [emigration], we see that they only occupied such stations temporarily." There is not a shadow of evidence to support the latter assumption of short occupation. Species that reside at high elevations on mountains have necessarily a somewhat restricted area of distribution, and would be more quickly exterminated in such localities than in lower and wider areas; for, as Dr. Wallace himself suggests, the raising of the snow-line, due to glacial causes, would not only reduce their area of distribution, but ultimately cause their extinction even in the very centres of their dispersal.

With regard to the actual route which this Inter-polar flora followed, north or south to either Polar continent, nothing need be said; but we cannot accept Dr. Wallace's conclusions, that when the South Polar continent became glaciated "these plants would be crowded towards the outer margins of the Antarctic land and its islands, and some of them would find their way across the sea to such countries as offered on their mountain summits suitable cool stations; and as this process of alternately receiving plants from Chile and Fuegia, and transmitting them in all directions from the central Antarctic land, may have been repeated several times during the Tertiary period, we have no difficulty in understanding the general community between the

European and Antarctic plants found in all south temperate lands." Now it is impossible to understand how any area could become "crowded" with plants (or animals) that are retreating from their normal habitat and entering areas where more or less adverse conditions of existence must prevail. Is it not more philosophical to assume that these mountain bases were the points from which the flora started, and that they are the bases on which the remnants of that flora will be preserved, whilst that portion occupying areas beyond them will perish through an adverse change of climate? Botanists (and zoologists) are too apt to overlook the fact that no matter how easily a seed (or in the case of an animal, an emigrant) may be transmitted from one region to another, it is perfectly useless as a colonizing or range extending medium, if the region entered is not adapted to its requirements and successful propagation. Breeding conditions must therefore always determine and control range extension.

We have precisely the same phenomena on the mountains of Asia—equatorial range bases and centres of dispersal of that flora which has spread north and south into the Arctic regions on the one hand, and into Australia, New Zealand, and Antarctica on the other. Just as we found to be the case in birds, the highest ranging species are the most widely dispersed; in the more temperate ranging types the identity is only of a generic value. This brings us to the consideration of such species as we have already described as "Inter-hemisphere." They are species that have ranged north or south from an equatorial base into the Temperate zones only of either hemisphere. The much greater difference of conditions prevailing in the two Temperate zones of the earth, than in the entire Polar zones (either by altitude or latitude), is reflected in the species occupying them. In Polar zones conditions are very similar throughout,—hence we find (especially as regards floras) various forms preserving specific identity throughout vast areas; in Temperate zones conditions are almost endlessly varied, so that species have been modified and multiplied in conforming to such varying conditions, and the equatorial range base is very frequently only a generic one. Thus we find in the floras of Australia and Europe that species of a Temperate zone in Europe are represented in Australia by very distinct species. Both must have sprung from a common equatorial range base (Borneo, the Moluccas, and New Guinea), one portion of a species emigrating as far north as Europe, the other portion as far south as Australia. The conditions which each set of individuals experienced were totally different, and modification of a specific value was produced, although in many cases they have both managed to retain their generic affinity.

When tested by the distribution of plants in Africa these facts assume even greater significance and suggestiveness. According to Dr. Wallace, there are no less than 60 genera of North Temperate plants in South Africa, none of which occur in Australia, and but few of the species characteristic of Australia, New Zealand, and Fuegia are found there. South Africa is now isolated from all the great southern land masses, and appears to have been so for a comparatively long period, a fact which has not only arrested the southern emigration of plants, but caused much modification. The Temperate flora of South Africa has been derived by southern emigration from range bases on the highlands and mountains of the equatorial regions, just as Europe has been so invaded by a northern emigration. But the southern movement has to a very great extent been arrested, because South Africa now contains scarcely any area which can fairly be classed as belonging to a Temperate zone; and this fact explains why the floral links with Europe are only of a *generic* character. On the other hand, nothing arrested the southern extension into Temperate climates in Australia, New Zealand, and South America from common equatorial bases, where the Temperate zone extends almost to the same (if more contracted) limits as it does in the Northern Hemisphere; and as a natural consequence many of the floral links between Australia and Europe, and even between America and Australia, are of a *specific* character. Dr. Wallace asserts that this phenomenon is clearly due to a *northern* emigration from an Antarctic centre of dispersal, which is obviously erroneous. Were South Africa prolonged at the present time as far south as say S. lat. 50°, so that it could present for occupation an equally extensive South Temperate zone as South America or New Zealand, there can be little doubt that no such differences would exist. The comparatively few (and presumably very ancient) resemblances between the floras of South Africa, Australia, New Zealand, and temperate South America cannot be remnants, as Dr. Wallace suggests, of an ancient vegetation once spread over the Northern Hemisphere, and driven *southwards* by pressure of more specialized types into these isolated areas, but are relics unquestionably of a dominant flora which started from an equatorial base, and ranged far south, not necessarily over continuous land masses south of Africa, although the probability is that such may have been the case.

We have precisely the same phenomenon among birds as among plants—Northern and Southern groups which are strictly confined either to the Northern or the Southern Hemisphere—north or south of the Equator, which must have started from an equatorial range base and spread north or south towards either Pole. Both Professor Huxley and Professor Parker have described the dominance of these Northern and Southern groups. Of course many groups may attain their highest development at remote and varying distances from this dividing line, or be common to both sides, or even become temperate or polar with little or no trace remaining of their equatorial origin from remote ancestral forms. As may be readily surmised, this new Law of Dispersal demands a continuous land connection equatorially. But this need not have been a synchronous one. Indeed the distribution of some groups of birds absolutely demonstrates that such was not the case. The elevation necessary to restore Antarctica—say 2,500 fathoms—would also be sufficient to connect the great land masses of the globe equatorially, and to a very great extent in the North Pacific and North Atlantic oceans. But such an elevation again need not have been synchronous; indeed the probabilities are that the Northern Hemisphere continents were united at a much more recent date than those of the Southern Hemisphere.

If this Law of Dispersal be true, it will explain the absence of southern types from the Northern Hemisphere, what Dr. Wallace aptly describes as "the singular want of reciprocity in the migrations of northern and southern types of vegetation." This Flora is a dominant southern one—a flora that has almost entirely spread from a base *south* of the Equator. One or two extreme northern outliers of this southern flora are to be met with on the Equator, and from such a base a few others have emigrated as far north as California, India, China, and the Philippines; but these are in every case exceptional, and are species or genera that should properly be excluded from that dominant southern flora. Dr. Wallace attempts to account for the "curious inability" of this southern flora to penetrate into the Northern Hemisphere by the totally different distribution of land in the two hemispheres; but dispersal, so long as it is normal, will overcome such difficulties, or at least overcome them to such an extent as to show some signs of the intrusion of organisms into adjoining areas, as we have already had abundant testimony. We can explain this absence of southern types from the Northern Hemisphere by one way, and one way alone, namely, that they were developed from a range base south of the Equator, and that the Law of Dispersal inexorably keeps those types to the Southern Hemisphere, and will continue to do so notwithstanding a change to adverse conditions which could have one effect only, their total extermination, just as we have seen to be the case with a once dominant northern fauna and flora. If we take the flora of a Northern zone, say between the 10th and 40th parallels of North Latitude, and compare it with a Southern zone of equal limits, say between the 10th and 40th parallels of South Latitude, we shall find the same inability of a dominant northern flora to establish itself in the south;

or, in other words, that the flora characteristic of that one in Europe, Asia, and Africa has not penetrated to South Africa or Australia, no more than the flora most characteristic of those countries has penetrated to South Europe, South Asia, and North Africa.

If this Law of Geographical Distribution be true, Polar dispersal of species—or in other words from the direction of the Poles towards the Equator—is a myth. To my mind we have overwhelming evidence to suggest that the grand centre of Life's dispersal across the globe is an equatorial one; and that from those regions where the greatest stability of climate, and the most favourable conditions for the development of animal and vegetable forms are to be found, Life in two grand streams has flowed Pole-wards. Glacial Epochs at either Pole have wrecked and exterminated all living things within their baneful influence, but on the return of more genial climatic conditions, Life in its endless forms just as often has spread northwards again (in the Northern Hemisphere) vigorous and strong from more southern bases, to repopulate the ice-freed lands with a fauna and flora adapting themselves to the diverse conditions of existence. I freely admit that these vast Polar lands, under the long-enduring favourable conditions that palæontological evidence compels us to admit once prevailed therein, have produced magnificent faunæ and floræ of high development, yet the Glacial Epochs have just as surely exterminated them, the only surviving relics being those that were able to maintain themselves in regions beyond such glacial influences, and where they MUST have been established before the adverse climates developed.

The recent discovery of sub-fossil remains of an extinct hippopotamus in Madagascar[1] is an event of profound and far-reaching importance. The presence of this huge mammal in Madagascar clearly proves that the southern emigration of the higher forms of life into the Ethiopian region could never have taken place at the date Dr. Wallace ascribes to it, and further proves that these large mammalia must have had an equatorial or southern base to survive the climatic vicissitudes of the Ice Age, as I have already insisted. As Dr. Wallace writes (*Island Life*, p. 448), "the most striking and characteristic groups of animals now inhabiting Africa are entirely wanting in Madagascar. Let us first deal with this fact, of the absence of so many of the most dominant African groups. The explanation of this deficiency is by no means difficult, for the rich deposits of fossil mammals of Miocene or Pliocene Age in France, Germany, Greece, and North-west India, have demonstrated the fact that all the great African mammals then inhabited Europe and temperate Asia. We also know that a little earlier (in Eocene times) tropical Africa was cut off from Europe and Asia by a sea stretching from the Atlantic to the Bay of Bengal, at which time Africa must have formed a detached island-continent such as Australia is now, and probably, like it, very poor in the higher forms of life. Coupling these two facts, the inference seems clear, that all the higher types of mammalia were developed in the great Euro-Asiatic continent (which then included Northern Africa), and that they only migrated into tropical Africa when the two continents

[1] Conf. Heilprin, *Geographical and Geological Distrib. of Animals*, p. 373; and *Nature*, January 24, 1895, p. 311.

became united by the upheaval of the sea-bottom, probably in the latter portion of the Miocene or early in the Pliocene period. It is clear, therefore, that if Madagascar had once formed part of Africa, but had been separated from it before Africa was united to Europe and Asia, it would not contain any of those kinds of animals which then first entered the country. But, besides the African mammals, we know that some birds now confined to Africa then inhabited Europe, and we may therefore fairly assume that all the more important groups of birds, reptiles, and insects, now abundant in Africa, but absent from Madagascar, formed no part of the original African fauna, but entered the country only after it was joined to Europe and Asia." Now in my opinion this is an entirely wrong interpretation of the facts as tested by our Law of Dispersal. Miocene (or any other) Emigration into Africa from the North, as described by Professor Huxley and by Dr. Wallace, could never have taken place. One can almost venture to anticipate the ultimate discovery of palæontological evidence of the former existence of these large mammals, even in isolated Australia.

The facts brought forward in the present volume place us in a position to understand more fully the futility of invoking a vast Antarctic continent to explain the various apparent anomalies of distribution presented in the Southern Hemisphere.[1] The restoration of this submerged or ice-clad Antarctic land mass can never be of such service in explaining the occurrence of closely allied forms in the now widely separated lands of the

[1] Conf. *Fortnightly Review*, February 1894, p. 194, and April 1895, p. 640.

Southern Hemisphere, as those naturalists that so strenuously assert its former existence appear to believe. The fauna and flora of Antarctica must have been exterminated with the destruction of that area—continent and inhabitants alike perishing together; the only forms, nay the only types, that would survive being those that chanced to have Northern Bases beyond the devastating influences which have succeeded in destroying pretty well half the sedentary population of the earth. Antarctica can therefore never explain the dispersal of species and types in now wide distant areas of the Southern Hemisphere; for not a single relic of the drowned and lost South Polar continent has been preserved to us by retreat from the slowly threatening doom, or by a Northern Emigration contrary to the inexorable Law of Life's dispersal. Such widely dispersed forms in many cases obviously of common origin only indicate the bases and the sources from which that doomed South Polar land has derived its inhabitants, from a previously much more continuous, more northern, or even equatorial base. (Conf. p. 57.)

The distribution of Life suggests to me the following conclusions. Firstly, where the land masses are greatest south of the Equator we should expect to find, and do find, the most important and extensive assemblages of species and types presenting the greatest amount of differences from such assemblages of types and species dwelling on the land masses north of the Equator; and these assemblages will to a great extent be homogeneous or otherwise in proportion to longitudinal continuity equatorially[1] of such Southern Hemisphere areas. Secondly,

[1] I use the term "equatorially" more especially in contradistinction to Polar.

where the land masses are greatest north of the Equator, we should similarly expect to find, and do find, the most important and extensive assemblages of species and types presenting the greatest amount of differences from such assemblages of types and species dwelling on the land masses south of the Equator; and these assemblages will to a great extent be homogeneous or otherwise in proportion to longitudinal continuity equatorially of such Northern Hemisphere areas. To a great extent this is quite irrespective of existing equatorial land areas connecting the Northern and Southern Hemispheres respectively.

In my opinion we have hitherto attacked the problem of migration and geographical dispersal at the wrong end. We have regarded Glacial Epochs, climatic changes, and physical mutations, as grand distributors of species, compelling Southern Emigration in one hemisphere, Northern Emigration in the other, rather than as vast exterminating influences which have repeatedly cleared one dominant flora and fauna after another from all the areas affected, the only surviving forms being those whose bases or areas in which Reproduction took place were beyond the fatal limits of glacial or other influence. The enormous scope for evolution in the oft-repeated march of life from the tropics or temperate zones towards the Poles combined with physical change must be apparent to every reader. Such periodical emigrations to a great extent account not only for the origin, but for the geological sequence of species. This Law of Dispersal will explain why in past ages we meet with a great extension of range of forms which are now limited to small areas—eloquent testimony to their inability to escape from adverse conditions, the Law of Dispersal forbidding range extension in perhaps the only way that safety may have existed.

With so many convincing facts before us, is it too much to assert that the fascinating science of Geographical Distribution, or the Dispersal of Life, will have to be reformed, remodelled on an entirely new basis, before we can hope to arrive at any uniform or correct interpretation of the phenomenon as it now exists?

We are aware of the wonderful powers possessed by most organisms to extend their area of dispersal—the marvellous contrivances of plants, the agency of flight in bats, birds, and insects, the swimming powers of fish and other aquatic creatures, all able, one would think, to emigrate this way and that, as fancy chose or necessity demanded. But are we to believe that all this range dispersal is purely of a fortuitous character; that accident controls its direction, and that chance shapes its course? I, for one, cannot bring myself to believe it, and am compelled to regard this mazy wandering and endless peregrination of Life across the globe as being inexorably subservient to Law.

INDEX.

THE END.

Richard Clay & Sons, Limited, London & Bungay.

www.ingramcontent.com/pod-product-compliance
Lightning Source LLC
Chambersburg PA
CBHW031848220426
43663CB00006B/541
* 9 7 8 3 3 3 7 2 7 2 5 8 6 *